MUDROS,
1915

*Two French nurses with the
Army of the East*

—

MUDROS, 1915

—

JEANNE ANTELME
and ELISABETH JARDIN

Translated by

BERNARD *de* BROGLIO

LITTLE GULLY PUBLISHING

2026

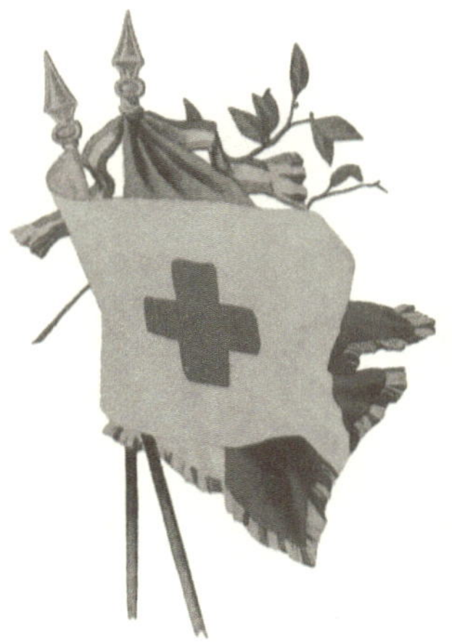

Originally published as—

Avec l'armée d'Orient : notes d'une infirmière à Moudros
(Paris: Émile-Paul Frères, 1916)

*Un Hôpital d'Évacuation aux Dardanelles: L'H.O.E. 1
de Moudros (Avril 1915–Février 1916)*
(Paris: Imprimerie de la Faculté de Médecine, 1920)

English-language edition by Little Gully Publishing, 2026
Translation © 2026 Bernard de Broglio

ISBN 978-1-7640773-4-7 (paperback)
ISBN 978-1-7640773-5-4 (ebook)

Little Gully Publishing
littlegully.com

A catalogue record for this book is available from the National Library of Australia

CONTENTS

This volume brings together two works by French women who served at Evacuation Hospital no. 1, Mudros, during the Gallipoli Campaign of 1915. Book One is a literary memoir by Jeanne Antelme, a volunteer nurse; Book Two, a medical thesis by Elisabeth Jardin, a student doctor. One captures the experience of the campaign through the eye of a writer; the other records the clinical reality of the hospital. Both are published here in English for the first time.

———

BOOK ONE

With the Army of the East

—

NOTES BY A NURSE AT MUDROS

—

JEANNE ANTELME

JEANNE ANTELME

———

AVEC L'ARMÉE D'ORIENT

———

NOTES D'UNE INFIRMIÈRE

A MOUDROS

PARIS
ÉMILE-PAUL FRÈRES, ÉDITEURS
100, RUE DU FAUBOURG-SAINT-HONORÉ, 100
PLACE BEAUVAU

———

1916

TRANSLATOR'S NOTE

This book has long intrigued me. The author volunteered as an army nurse and accompanied the French expeditionary corps to the Dardanelles in 1915. Hers is the only literary account by a woman who served in the campaign, distinct from the many no-less-important letters and diaries by Australian, Canadian and British women who served in field hospitals and on hospital ships during the Gallipoli Campaign.

Jeanne Antelme was Mauritian-born, of French heritage, but the child of a British dominion—France having surrendered the island to Britain in 1810. Her fervour for Gallica chimes with that of the translator, whose *grand-père* was born in Mahébourg, on the south-east coast of Mauritius, overlooking the bay where French ships won a famous victory over the Royal Navy.[1] He passed on to me that irremediable nostalgia that colonials have for their distant and departed mother. (Since childhood, I have mimicked him by keeping a bust of Napoleon Bonaparte on my bookshelf.)

Jeanne Louise Judith Antelme was born in Port Louis, Mauritius, on February 9, 1883, the daughter

1 The Battle of Grand Port is the only naval victory inscribed on the Arc de Triomphe in Paris.

of Eugénie Couture and Louis Edgar Antelme. Her father was a scion of the island's powerful sugar oligarchy—the son of the legendary Sir Célicourt Antelme, a planter and politician who dominated the colonial legislature for decades. Her upbringing within this influential Franco-Mauritian family placed her at the pinnacle of colonial society, with life divided between Mauritius and Paris.

On August 21, 1909, at Criel-sur-Mer on the coast of Normandy, Jeanne married André Noblemaire. It was a brilliant match that propelled her into the highest circles of the *haute bourgeoisie*. André was the Deputy Director of the *Wagons-Lits* company (operators of the Orient Express) and the son of the Director General of the P.L.M. (*Paris-Lyon-Méditerranée*). As the premier railway network linking the capital to the Riviera, the P.L.M. was a pillar of the French economy, and the marriage established Jeanne firmly within the country's elite.

Her literary debut arrived in July 1914, on the very eve of the conflict that would define her oeuvre. Titled *Vivre* (To Live), the volume was released under her maiden name—a significant choice after five years of marriage—and bore the imprint of Alphonse Lemerre, the illustrious publishing house of the Parnassian poets and Paul Verlaine. Far from a sedate society memoir, the work was a collection of 'bold and original' chronicles, traversing landscapes both geographic and emotional: from travelogues of the Seychelles and the Indian Ocean to meditations on 'Desire' and 'Madness.' A contemporary review in *Les Nouvelles* betrays the masculine unease of the time; while the critic admired her 'precise and colourful' style—comparing her realism to the Naturalist school—he was visibly unsettled by her 'feminist ardour.' Defining woman as a being 'weak by definition,' he recoiled at Antelme's dissection of domesticity, complaining that she portrayed the traditional head of the family not as a guide, but as an 'executioner leading his own to the slaughterhouse.'[1]

Germany declared war on August 3, 1914, and Antelme's second book—*Soldats de France: simples esquisses* (Soldiers of France: simple sketches)—moved from the domestic battleground to the literal front. Published by Delagrave in 1915, the volume featured a foreword by Baron Guillaume, the Belgian Minister in Paris, and attracted royal attention; the illustrated daily *Excelsior* reported that both the Queen of England and Queen Alexandra had sent their 'warm congratulations' to the author.[2]

Writing in *La Revue hebdomadaire*, a critic praised the text for bridging the gap between the trenches and the home front—'a soldier sees the war at too close a range; a mere civilian, from too great a distance'—a feat achieved through Antelme's 'passionate zeal for documentation' and 'unique access' (likely facilitated by her husband's

---

1 *Les Nouvelles* (Paris), 30 July 1914, p. 3.

2 *Excelsior* (Paris), 2 June 1915, p. 10.

position and the service of her brother, Fernand Antelme, in the French army). The sketches range from abstract observations—capturing the atmosphere of mobilisation, describing Paris in wartime, a definition of the 'Fatherland'—to intimate vignettes: seventeen-year-old Savoyard conscripts at a railway station, the 'white bursting of shells' on the dunes amidst French marines and North African troops, and transcriptions of soldiers' letters. Explicitly placing the work 'under the aegis of charity,' Antelme dedicated the proceeds to French and Belgian war victims, yet the review concluded with a characteristic period distinction: that she had treated the subject 'not as a writer, but as a woman, with all her sensibility.'[1]

In August 1915, aged thirty-two, Antelme sailed east to the Aegean island of Lemnos as a volunteer for the *Société de secours aux blessés militaires* (SSBM), the oldest of France's Red Cross societies. Her destination was the rear base for the Gallipoli campaign, where French forces were fighting a brutal war of attrition alongside the British at Helles—a contribution often overlooked in Anglophone accounts. Alongside a small team of volunteer nurses, Antelme joined the staff of *Hôpital d'évacuation* (Evacuation Hospital) no. 1, a sprawling facility perched on a barren hillside just south of Mudros town. Working under head

---

1 *La Revue hebdomadaire: romans, histoire, voyages* (Paris), 16 October 1915, p. 397.

nurse Yolande Oberkampf, Antelme served in what became the single largest hospital on the island; with a capacity exceeding 1,500 beds, it dwarfed nearby British and Dominion units.

Antelme's direct, first-hand experience of the horrors and immense logistical challenges of military medicine in this campaign formed the basis of this work. *Avec l'armée d'Orient: notes d'une infirmière à Moudros* (With the Army of the East: notes of a nurse at Mudros) was published in 1916 and awarded the Jules Favre Prize by the *Académie française*—a biennial distinction reserved for literary works by women. As the ancient and official guardian of the French language, the Academy's endorsement signalled that Antelme's notes were more than a diary; they were a text of national substance.

Antelme also lent her voice to the periodical press, where she illuminated the often-obscured theatres of the war. In columns such as *'Nos Marins'* (Our Sailors) for *L'Echo de France*,[1] she championed the 'invisible' vigilance of the fleet, capturing the tension of the submarine war through the 'perpetual vagabondage' of patrols. Her essay in *L'Information* moved from the moral to the visceral. In *'Conscience et Devoir'* (Conscience and Duty), she analyzed the 'fierce coquetry' of British patriotism—noting how the ardour of the suffragettes had been placed at the service of country—while challenging French

---

1 *L'Echo de France* (Paris), 9 October 1916, p. 2.

women to equal stoicism.[1] Yet her most arresting contribution remained the personal: the February 1917 account of her brother Fernand's captivity. Eschewing literary artifice for raw testimony, she chronicled his humiliation in chains at Nuremberg and his suffocating escape beneath crates in a sealed railway wagon, transforming a private family drama into a public narrative of resistance.[2]

Yet, by the spring of 1919, having returned to civilian life, Antelme found herself the subject of sensational headlines in the Parisian press regarding her divorce from André Noblemaire. On April 29, outraged by a judgment rendered against her, she marched into the office of the Deputy Public Prosecutor, *Monsieur* Boutigny, and attempted to read a prepared statement. It was only when the magistrate refused to listen to her 'long factum,' declaring the matter closed, that she seized an electric lamp from his desk and hurled it at his head. The magistrate escaped with a cut thumb, but Antelme was briefly incarcerated in Saint-Lazare prison.

After this *succès de scandale*, the trail goes cold. Our author largely disappears from the historical record, though faint traces suggest she traded her pen for the visual arts. The catalogue for the 1923 *Salon d'Automne* identifies a 'Mme Jeanne Antelme',

1 *L'Information: financière, économique et politique* (Paris), 3 January 1918, p. 4

2 '*Histoire d'un évadé d'Allemagne*', *La Renaissance: politique, littéraire et artistique* (Paris), 17 February 1917, pp. 2-4.

categorised explicitly as *'Mauritienne (britannique),'* living at the fashionable 33 rue du 4-Septembre. At the celebrated annual showcase for modern art and the avant-garde, she displayed two paintings: *Fillette* (Little Girl) and *Fleurs* (Flowers). A year later, the *Journal Officiel* lists her as a sculptor claiming *droit de suite* (resale royalties).[1] Her later life is not documented; that story remains to be told.

What, then, of the text that anchors this translation?

Military historians may hesitate over Antelme's lyrical renderings of the landscape, yet terrain is fundamental to understanding this campaign, and the environment she describes was a constant preoccupation in soldiers' letters. Her sensory observations—of light, dust, sound and sea—are not mere ornamentation; they provide a vital texture to the military sphere. Indeed, the work abounds in specific logistical details: the laboratory in the cellars of Sedd-el-Bahr fort, the first-hand accounts from the battleship *Suffren*, and the makeshift medical infrastructure on Lemnos.

Similarly, the religious passages may sit uneasily with the agnostic reader, yet one must recall that at this time, faith was not relegated to the margins. The French medical services were, in large part, a vocation populated by seminarians and men and women of the church—including those religious

---

1 *Journal officiel de la République française / Lois et décrets* (Paris), 25 August 1924, p. 7925.

exiles who returned to serve the Republic that had once banished them.

Finally, the author's intense patriotism may be difficult to reconcile with current sensibilities, where love of country is so often conflated with nationalism. We must remember that these pages were penned when the invader stood on French soil. The war's outcome was a matter of grave doubt, and the nation's very existence hung in the balance.

If the translator's bias has been laid bare, it is with the hope that the reader will find similar value in this overlooked text.

* * *

The photographs that illustrate the text are drawn from a number of sources, but the most valuable are those of Elisabeth Jardin, who served alongside Antelme at Mudros. I am indebted to Bill Sellars, with whom I acquired parts of the Jardin collection; he has generously shared these images that offer a new perspective on Mudros, 1915.

For their knowledge, support and friendship, I gratefully acknowledge Margaret Carter, Spiros Hrambanis, Liz Kaydos, Steve Moore, Markos Psarakis, and Serpil and Bill Sellars. Above all, I must thank Cheryl Ward, author of the play *Through These Lines*, who took me to Lemnos in 2011 and kindled an interest that is far from being quenched.

B. dB.

DEDICATION

I dedicate this book to those of my own people who served France in their own way. To my maternal great-grand-uncle, *Commandant* Jacques Couture, *Chevalier* of the Legion of Honour. Wounded in action at Charleroi (6 June 1794), at the Battle of Hohenlinden (3 September 1800), at the Battle of Eylau (8 February 1807), in action at Eckmühl (21 April 1809) and on 9 February 1814 at the outposts of Harburg while commanding four companies of the 1st Battalion. Mentioned several times in army corps despatches, he received letters of commendation from the Marshal of the Empire, Prince of Eckmühl, for his conduct and the sangfroid he displayed in various actions during the blockade of Hamburg: particularly those of 10 September 1813, and 20–22 January, 9 February and 12 March 1814 at Harburg.

To my maternal great-grandfathers, both gallant officers of the *Grande Armée* and mentioned several times in army despatches. To my paternal great-grandfathers, who contributed to the defence of Mauritius in 1815, both on land and at sea. To my paternal grandfather, Sir Célicourt Antelme, K.C.M.G., *Chevalier* of the Legion of Honour, who always proudly asserted his French origin and bid a memorable farewell to the French language when it ceased to be the language of the courts in Mauritius.

To my father, the Honourable Edgard Antelme, who, in the various voyages of exploration he undertook in 1873, 1874 and 1875 to Australia, South Africa and Madagascar, strove to foster a love for France.

To my brother who, enlisting in the regular French army at the very outbreak of war, was wounded and taken prisoner in September 1914 after distinguished conduct, and in March 1915 was sentenced to eleven months' fortress confinement for attempting to escape.

J.A.

CONTENTS

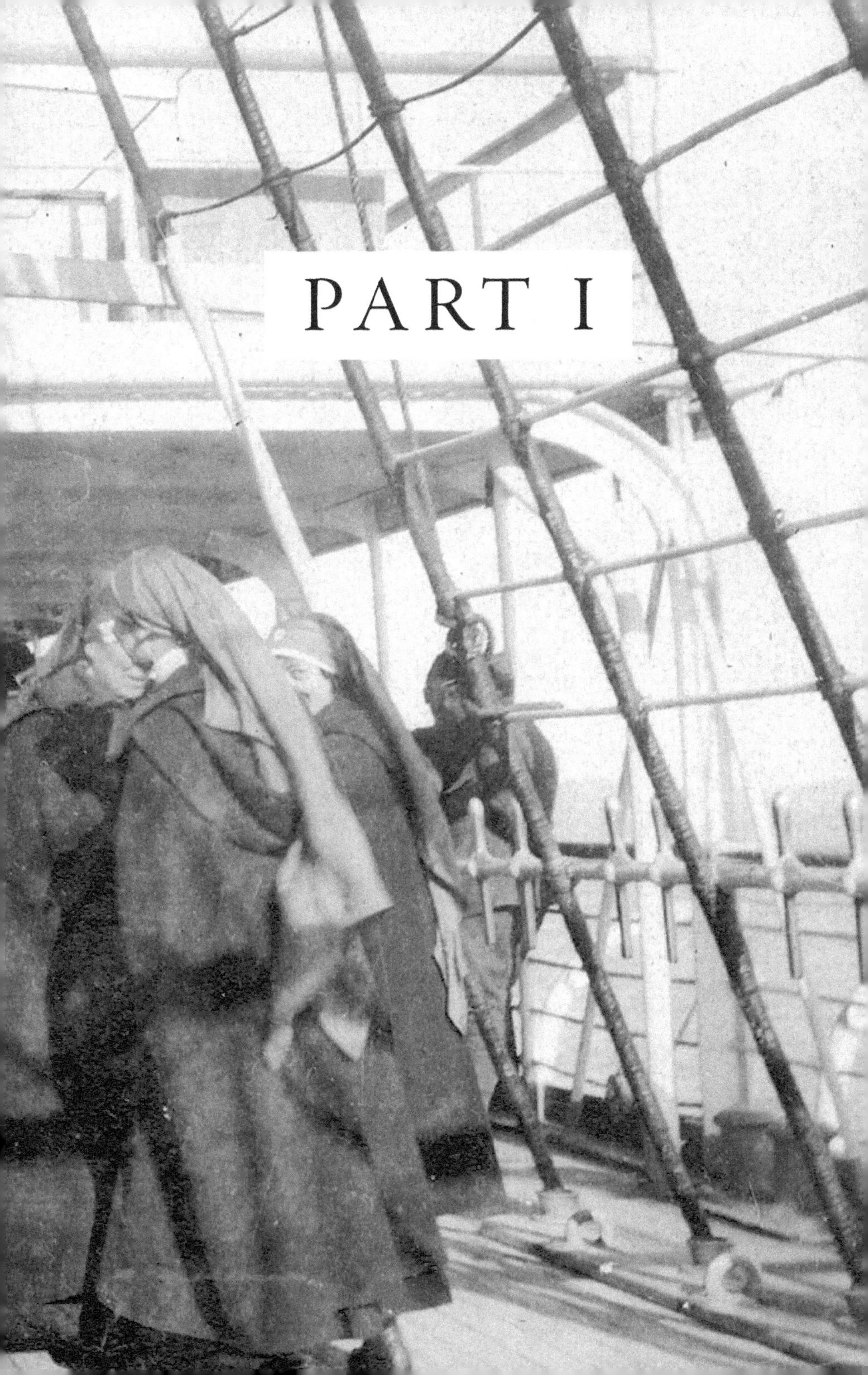
PART I

Torpedo boat, Lemnos, 1915.

(Lieut. Homolle, Ministère de la Culture / APOR 051221)

I

Mudros, August 1915.

A torpedo boat!... Some distance astern. A torpedo boat, unsuspected in these waters... Oh! She flies over the waves, hurtling forward, half-hidden by the spray gushing from her sides... A swirl of foam, then a manoeuvre of boldness, even elegance... Hugging the flank of the hospital ship, she badgers the vessel with questions... "Who are you? Where are you from?" the megaphone barks... The words lash out...

Our eyes linger on the slender grey hull. We look not merely from habit; it is more than that... We look solemnly, and—it must be said—with emotion, at this handful of men, these sailors fighting in their own way. To police the seas, to keep watch without cease... always to keep watch; it is no easy task...

And then, as minutes slip by, an unspeakable sadness, a frightful melancholy takes hold—a sadness that seems destined to last forever, so stubbornly does it cling... Is it the weight of days lived, the days of yesterday and the day before?... Is it this present hour, this unknown we are approaching?

Away with conjecture! One must, if one must... One must steel oneself, shake oneself; one must breathe deep, head held high, and leave much of oneself far behind... One must be silent, clutch with both hands the heart that sometimes aches... One must trample one's pain underfoot.

One must! ... The mine barrages... That great rosary of red buoys that closes the roadstead... The bay... very wide... important...

This evening, a light, fine and very gentle, hangs in the air... Blues and mauves play upon the rigging, on the ship's prow, between the wavelets, and cloak the surrounding mountains whose graceful outlines spread in radiance. It is luminous, clear, so very limpid. The immense gravity of summer twilights...

And now something anomalous, never seen before, unsuspected... Countless ships crowded together, massed in shelter... Great and small, they are entangled, masts interlaced... Obsolete funnels, modern stacks. Monitors and battleships, silhouettes of old frigates, elegant ultra-modern torpedo boats, cargo boats sweating and swearing alongside the white hulls of hospital ships, tiny submarines brushing past some gigantic *Mauretania*,[1] fishing boats harking back to gentle caravels, Greek caiques, wheezing tugs, launches and trawlers, all motionless... as if petrified.

1   British ocean liner, then the fastest ship in the world and one of the largest. She served as a troop transport and hospital ship during the Gallipoli Campaign.

It is the hour of twilight. The most precious hour of all, the hour of prayer and contemplation, the tranquil hour... the hour when the soul finds release... understands, and rises to the infinite... the divine hour...

Sunsets of the East, so magnificently sung of, was it thus you were dreamt of? Were you dreamt of to soothe the peculiar thrill of those who would give everything for their country, yet can do so little...

Why do tears fill certain eyes to-night? ...

A great silence reigns... And it is like a prayer rising towards Heaven...

And still the ships, like great black birds, lie motionless... upon the motionless water...

But—a surprise! Funnels glide by, lights appear. One must watch, without moving, for fear of disturbing the silence, for fear too of being mistaken...

But... but yes, that is it: among the black birds, other black birds are moving... many black birds, gathering now from all sides, seeming to obey mechanically... An order, perhaps?... Being executed in unison? They glide without a jar, without the slightest hesitation; they thread their way, solemn and silent...

They passed, all lit up as if for a gala... And night came down, a night like many Eastern nights, a very luminous night...

Oh! Those lights that were so close... the lights of those ships filing past—they vanished in a flash, very abruptly.

Of course!

"All lights out..." The standing order... One must navigate with eyes fixed more than ever on the compass... on the sky, sometimes, to seek the stars that serve as landmarks... Songs... cries... good luck wishes... A transport passes... On deck, action stations... A human swarm from which bursts that admirable enthusiasm which has not aged a day... And then, in its turn... the transport has vanished into the night...

Silently, as if with measured steps, like a nun walking in church, a ship glides almost religiously in the wake of the other, the transport... A ship not so very large, a ship all white... A broad luminous green band... and a large red cross, luminous too... Oh! She passes without a cry... without a sound... She is almost lifeless. And she will not extinguish her lights beyond the mine barrage: she must not; 'on account of a convention', she never extinguishes them... She is only a little sorrowful... melancholy... The waves play with her cross of fire which scatters across them... She continues on her way, calm and resolute. From afar, from very far away, she remains visible. Her cross is now but a great ruby, blood-red... And every head has bowed... Make way for those who have fought...

On the hillsides where the camps clamber up, fires are lighting. The routine of evening life resumes... Life far from France, in a setting unknown back home... under an Eastern sky, a night of war and unshakeable courage...

Oh, let us pray, let us pray to God, always...

———

View over the French hospital to the
harbour and town of Mudros.

(Elisabeth Jardin Collection)

II

Mudros, August 1915.

Oh! This dust that blinds us, that makes us turn our backs... Eyes burn and mouths are filled with it. It grinds between the teeth. It works its way down your neck, slips between your shoulders.

It is a day of excessive heat and, as the wind blows, the dust dances a frenzied jig. Jostled, lifted up, it rises, it climbs high, it hurls itself upon you, lashes you like a fury. And wherever one goes, under the bell tents or into the huts, it pursues you, haunts you...

Oh, this dust! I wager you have never seen its like. Just think: the roadstead itself has disappeared beneath the yellow cloud clinging to it. One can no longer see the ships. One sees only the great dusty pall that seems to insult the sky. These days of dust are the worst misery... It carries with it all the bad germs and is, in a way, a sowing of death...

Add to this the innumerable flies which, to escape it a little, fill the huts and tents. How terrible they are, these flies, and how they seem to bear you a malice! They are everywhere; they cover the slightest woodwork, the smallest scrap of canvas, anything that offers a surface. And they pack so tight together that one can no longer tell what they are resting on. They swarm, they drone, they rise in a mass to

23

pursue you, to torment you. They settle on your lips, on your eyes; they get into your mouth. Oh! They are intolerable... It is a veritable suffering...

I was forgetting to tell you of another insect, as odious as it is obstinate. Its very nature shocks me and I am almost ashamed to name it to you... I speak of fleas... Yes, fleas, as countless as the flies, and they too mount the assault... We caught them off each other while tending our patients or while chatting. Sometimes we took two or three at a time, and the gesture was so natural that we did not stop to say thank you. They climbed up one's neck. One felt them strolling about, not singly, but in company. Sometimes they stopped, seeming to choose their spot, then off they went again, up and down.

We had not a patch of skin free from bites. At night one felt them, still prowling. One was woken twenty times, and twenty times—uselessly, moreover—one tried to hunt them down... Wasted effort, they started up again at once... Every evening, before turning in, we carefully spread out our things. And we hunted for them, the nasty beasts... The most patient among us reached a tally of two hundred to two hundred and forty... And this, I assure you, is no joke...

To our credit, I must say that we resigned ourselves to it... But what was hardest for us was seeing our poor patients who had to struggle by night with the fleas and by day with the flies... Above all, the poor severe cases! You had to see those flies entering their mouths and emerging in a buzzing mass when one drove them off. Their eyes were quite filled with them. The typhoid cases held the record. It was no use putting up mosquito nets; it did nothing. Go and try to fight such an invasion. It was unthinkable... The flies were found everywhere... In the smallest vessels, covered or uncovered, for scarcely was the lid lifted than they threw themselves in blindly. I will not speak to you of the other patients, the dysentery cases, firstly because it would disgust you and then because it would hurt you... You see, when you find yourself complaining, think sometimes of those poor young French soldiers who, in the East, suffered more than any... And yet, God knows we tried to remedy those sufferings. But there are sufferings inherent to situations, and theirs, you know, was not exactly a cheerful one.

Add to that an absolute lack of water. And yet, here we were in August, when conditions had greatly improved... Two months earlier, the hospital had gone three days without making soup, and the patients held out their mugs with a desperate gesture...

Let me tell you, it was something of a vision of hell, that camp in high summer. You had to see those great emaciated skeletons burning with fever, thrusting out, in unconscious movement, legs and arms that were less than bone... All that under canvas, mingled with dust, flies, and fleas... Ah!

The smell that rose from it... One needed a strong soul, or a will of iron, not to flee from pity and horror... At that time, several died every day...

Towards three o'clock, the little carts would line up before the hut where the dead were laid. A picket rendered honours, the bugle sounded the salute, and then, beneath the Tricolour, the mules, one by one, bore away their burden...

Great plumes of dust escorted them... The wind followed hard upon... It was a cheerless sight...

We had between fifteen and eighteen hundred sick. Neither the doctors nor the orderlies spared their pains—ah, the fine devotion I witnessed there... Everyone looked after one another. Everyone supported one another... And if at times we assumed a somewhat fatalistic air, it was because a great pity wrung the heart, and one felt a sob waiting to break...

Ah! yes, our hospital, for all its misery, was still the best of the lot. The British themselves marvelled at it...

And yet, had you entered one of our tents, you would have seen, on those sagging, soiled straw mattresses, heaps of men shivering in the throes of death...

Our camp certainly occupied the best site on the whole island... It was immense. The Turkish prisoners' camp adjoined it. Then came the Zouaves' camp. We were perched on the hillside, with the whole roadstead at our feet... Opposite us, camps stretched away into the distance... always those bell tents, so delicately placed that they seemed like great flower petals...

I shall always remember that arrival at Mudros... The vision of horror I had received by day, seeing all that suffering, that aridity—not a single tree, not a single plant—that impression was transformed by evening...

The wind had dropped; a beautiful sunset welcomed us. One of those twilights that cannot be described. I had never seen the sky so beautiful... The stars came out one by one, barely distinct at first, while the sea turned an intense blue, and great pink swells played over the whole... With the silhouettes of the old Greek caiques in the foreground, some of which had unfurled their great, pure white sails, it was something undreamt of, something marvellous... Dominating the mountains opposite, I could make out the transfigured summit of Mount Athos, while to my right, Samothrace, with its deep blue spine, resembled a great luminous beast...

Nights such as these, you see, are great lessons; and as I kindled a large fire between two stones to heat a little water, strange thoughts came to me... Yes, I had to thank God for having brought me here. I would be of use—I hoped so. I would apply myself to it with all my strength. And the sight of all this suffering showed us the vanity of the self, and how far we must cast from us anything that was not the supreme laws of love... The sky, up there, was Eternity; it was the desire possessed by each of us; it was our soul... Here on earth was our body, with its demands and its fears... It was necessary to master that, to enjoy fully what was best within us... One had to ask nothing more of life, and take only what it gave us... How indulgent we must be towards those who do not know; how we must pity them... You see, beautiful nights have never brought me anything but good thoughts... And before my roaring fire, over which I watched carefully, I thanked God... The stars above were still climbing... And there was such tranquil gentleness in the air that I knelt down, gazing at the sky...

————

III

Mudros, August 1915.

Evening prayer at the camp is now an established custom... One is free to attend or not. It takes place as night begins, outside in the open air so that there is room for all... It is said on the hillside, amidst the stones, amidst the dust, under the open sky...

The setting is immense, vast. The camp lies below us, and other camps, whose clustered tents resemble a honeycomb. The harbour, crowded with ships, their lights motionless. The silhouettes of Mounts Khadia, Shako—what are their names?—where sometimes great moonbeams steal through!

It is all this that the eye takes in, at the hour when twilight ends.

And it is here that one comes to pray... in this isolated corner where the Tricolour covers the wooden crate serving as the high altar... with dented candlesticks, aged by wear, lanterns with glass panes spattered with candle-grease, all rusted, all rickety...

One goes to evening prayer, eyes full of dreams and gentleness, without fear. Men arrive one by one... Seated in a circle by the golden glimmer of a lantern, one looks for the hymns to be sung...

The officiant has taken the golden monstrance from its case... And now he proceeds with his own vesting... First the surplice, which he slips carefully over his soldier's greatcoat... now the stole... A strange combination...

The priest has knelt down. And everyone has taken their place, very grave... Heads are bowed... even those unseen, lost as they are in the night. There are faces which the light strikes full on: gaunt, sallow masks, yet resolute. There is dignity in the bearing, despite the worn-out uniforms...

Other faces less distinct, being further back. Eyes that alone seem alive, huge, feverish. Collars turned up to the ears to dull the shivering that makes teeth chatter... Standing, or seated on the stone embankments, on steps dug out of the very earth... The attitude is one of reverence...

Besides, it is easily explained... they have come here because they believe...

Zouave chechias; territorial Zouaves, prosperous and handsome in their impeccable dress; the rugged faces of old warriors, grave and serious. Wild-haired infantrymen, emaciated wounded. A Senegalese Spahi, too. A proud bearing, with enough disdain to justify his brilliant uniform... The light strikes him by half, and the shadow stretching behind makes him seem all the more fantastical. He sits, legs outstretched, calm, imperturbable... very attentive... The light from the altar strikes him full on: a singular illumination, a tableau such as Rembrandt might have dreamt. Silhouettes take shape in the shadows, faces indistinguishable, hands clasping in a soft, mechanical gesture...

How many voices—I no longer know—have struck up the *O Salutaris*. Voices soaring, full of breath, climbing, rising towards the heavens; men's voices, full of magnificent resonance... The hymns rise above the men... God!... God!... God!... God who fills hearts, the universe, and who bursts forth on every side... God who reigns, and whose strength and will are imprinted on every thing... God whom one would wish to worship on both knees, prostrate in the dust... God to whom one surrenders... God to whom one stretches out one's arms in an immense appeal...

Oh, that evening prayer!... that prayer said up there... in the open air... in common... All those men, combatants of yesterday and of to-morrow, in that gathering which effaces every social rank, which unites the poor with the rich...

One comes to prayer, to sit on that bench with its rickety legs... Beside one, some brave legionnaire, shy and awkward in his boyish sincerity. Who is unaware—it comes so naturally to him—of the meaning of those crosses, *Croix de Guerre*, *Médaille Militaire*, that he has pinned, somewhat haphazardly, to his chest... One loses oneself in this crowd of men who simply do their duty—out of habit, perhaps, but assuredly with grace...

He who has not seen... does not know... cannot know...

On their knees, all of them... All, foreheads bowed very low... Now the light glides over the heads, over bowed shoulders... All on their knees... Yes... A divine benediction...

The monstrance rises and falls, and the little bell rings... rings... in an immense silence... It rings while foreheads bow, ever closer to the earth... The benediction draws to a close... The monstrance, its gold gleaming, alone illumines the night... Silence...

Now! All in chorus...

Parce Domine... Parce... populo... tuo... Ne in æternum irascaris nobis... Parce... Domine...

Hear us, O Lord... Hear the anguish of our hearts... ready to burst, yearning for You...

Parce... Domine...

A thin green ribbon, a red cross—both luminous— glide yonder over the roadstead... in great mystery, in great silence, commanding silence...

Parce... Domine...

Forgive... Forgive us, O Lord... And then... And then?...

Oh! It is in a whisper one must tell You, in ardent supplication. Oh! God of light... preserve the mothers, watch over the orphans... soothe the countless sorrows...

Fires in the neighbouring camps light up one by one. Glowing red braziers, scattered or clustered close. And then, further off, far, far away... the great sinister pyre... where the Hindus burn their dead... And to-night, no doubt, some macabre ceremony is unfolding...

Hymns follow, hymns that everyone knows... Oh, they sing as if to soothe their own hearts...

The chaplain has drawn closer to the men; a soldier holds the lantern, shedding a light almost too brutal upon him who has charge of souls... His eyes move from one to another... He speaks... as though speaking to each individually... Farewell to ceremonial protocol... We are all family here...

"My brothers, I have but two words to say to you, I do not wish to keep you... Before we part this evening, think, I beg of you, that in this camp, to-night, many of your comrades are dying, others are dead... Close beside you, the death throes have begun... Fathers, mothers, wives and children will weep... Pray... Pray for them... Pray that God gives you the courage to endure all things... Pray for France, who alone must be—and shall be—victorious."

"Now, good evening, my friends, good night... and until to-morrow, if you are willing..."

Parce... Domine... Parce... populo... tuo... Ne in æternum... irascaris nobis... Parce... Domine...

The words and notes haunt you—resounding, singing in the ear, in the heart...

Stumbling, supporting one another, amidst the muffled clatter of heavy hobnailed boots and the indistinct murmur of hushed voices, lost in an intimate, soothing silence, the soldiers descend the rugged slope leading back to camp...

One must keep silent this evening, knowing how to hold God within oneself—God the Creator... God Eternal!...

Three tiny donkeys, graceful little wanderers like djinns, strike a few stones with their tiny hooves... They shake themselves, almost merry... It is lovely, and very gentle too... And it does not disturb the silence in the least...

———

IV

Mudros, August 1915.

To-night, our camp treated itself to the luxury of a magnificent sunset... Never before had I beheld the like. I was walking towards it, with—spread out before me—one of those flocks of sheep whose long white fleece trails to the ground... They are, I believe, unique in the world... They were so beautiful that I stopped. They were descending the slope, and their fine, graceful brown feet struck the hard earth almost soundlessly. It was only a great rustling. The shepherd stood motionless, gazing at the setting sun. His heavy cloak did not encumber his movements, and he resembled some statue seen I know not where...

Suddenly, as if springing up by magic, a great rosy light spread through the air... From pink it turned to red... My sheep now bore great red fleeces, all illuminated. It was delicate, soft, yet violent all at once... The flock, too, watched this sudden metamorphosis with large, gentle eyes. The sheep halted, bleating a plaintive note.

The earth turned red... Our camp was transformed. Our pale grey tents became true tropical flowers of magnificent crimson. Faces flushed pink, blossoming into beauty... The red trousers of the Zouaves intensified with light, the chechias shone resplendent. The hills followed suit...

Then came a great wave of mauve; everything on earth appeared like large autumn crocuses... The sheep, their fleeces beginning to glow, moved softly away, laden with light. And the slow bells cast into the air the very gentle note of a beautiful summer twilight...

―――――

View south-west from the hospital.
(Jean Savin, Ministère de la Culture / APOR035119)

'At the door of my ward.'
Evacuation Hospital no. 1, Mudros.

Elisabeth Jardin was a medical student, later doctor,
whose photograph album illustrates much of this book.

(Elisabeth Jardin Collection)

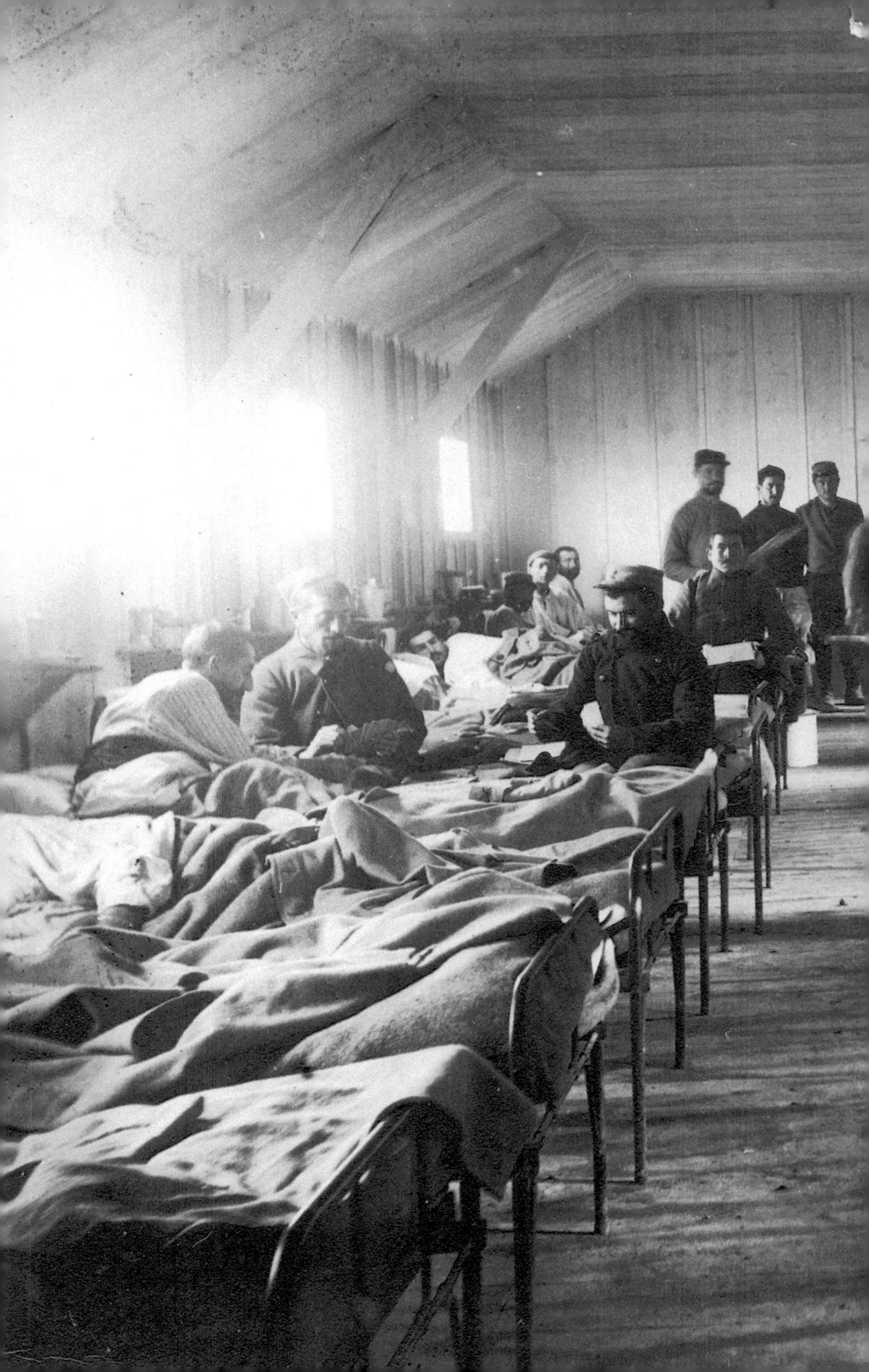

PART II

A burial at Mudros, photographed by a French soldier.
(Eugène Victor Marizy, Archives départementales de la Marne)

I

Mudros, September 1915.

To-day the funeral took place of one of the doctors in our unit.[1] Twenty-six years old!... After most brilliant conduct at Gallipoli, he had been sent here because he was seriously ill. He remained at our hospital, refusing to be invalided back to France. He had even resumed duty. All the patients loved him. He seemed robust; indeed, he was. A strapping great lad, sturdy and strong, who looked you straight in the eyes with honest frankness.

And then, in ten days, typhoid fever carried him off. You can picture it—the high fever that, in a few days, turns a living man unconscious, and then into a corpse... The hospital is in mourning. He had won every heart. And he was not the first to go this way... Ah! this typhoid, what misery it has wrought!

I remember sitting up with him one evening, though he was not in my ward. I recall those large blue eyes looking at me so kindly, so gently,

---

1 Perhaps *Medecin aide-major de 1ere classe* (Lieutenant) Sylvain (Jean Louis Sylvain) Dessaigne, 175th Infantry Regiment, who died of sickness (typhus) on 6 August 1915 at Mudros. Dessaigne was twice mentioned in despatches at Gallipoli for his gallantry in evacuating wounded from the frontline. His brother's name was Joseph.

while I replaced the cold compress on his burning forehead!... Ah! yes, you know, such death throes rend the soul; and feeling one's powerlessness to arrest destiny—that is what hurts the most... If one could but give something of oneself in those moments... If only one could!... But one can do nothing, and that 'nothing' weighs terribly heavy...

They had put him in a little room of bare wood, all alone. His poor iron bedstead seemed too small to hold that great frame... He still looked so healthy...

Yes, I remember... It remains engraved on my eyes, on my soul... The younger brother, who had wanted to follow the elder, the first-born, had managed to get sent to our part of the world. So, when he fell so ill, he came at once—the cadet... He never left his side... Kneeling by the bed, he laid his head on Jeannot's shoulder—as he called him... There were broken words, memories evoked and recounted. And clasping him tighter in his arms, he repeated:

"Oh! Jeannot, do you remember? Do you remember when we went to classes together? It was always you who prepared my lessons... say, Jeannot, do you remember?... No? You don't remember?..."

And he turned to me, sobbing, powerless... "You see, he hears nothing now, he knows nothing now..."

And he would begin again:

"Say, my little Jeannot, you're going to get well, and we'll go back to our happy little life..."

Ah! you know, it is better that I fall silent... For I myself no longer feel strong enough to go on...

And the younger brother went on... "Yes, the other evening, I remained alone with him for an hour... I needed something, I left him for scarcely a minute... When I returned, he was sitting up in bed. Frightened, I asked him what he was doing there, and he answered: 'I am waiting for Mother...' Those were his last words..."

I tell you, there is nothing sadder than deaths such as these...

He was borne to the cemetery with full honours... There was wind, and dust, and we were all spattered by it... Everyone was sorrowful, and many wept... She who, in concert with the doctors, had fought so fiercely to save his life, followed too, her heart utterly capsized... The bugler marching at the head of the procession cast great, deep, mournful notes into the sky every two minutes, sounds that seemed a reminder of the hour we were living... It was as if he were calling all attention to the dead man. And he seemed to say to you: Bow your heads... Salute... Understand. Ah! yes, my God, how those notes spoke... How he knew to modulate them... It was a great rallying cry that recounted, that said what had to be said...

As we passed the Greek well where women were drawing water, one of them stepped out from the group and cast upon the coffin a poor little flower, which had grown God knows where, and at the cost of what effort!...

———

Postcard of the French battleship *Suffren*.

II

Mudros, September 1915.

Our afternoons were often spent talking at length. Seated in a circle, the men perched on their kits or at the foot of their beds, the sickest among them buried beneath their sheets, I seated on some packing case or on a rickety old bench that was none too strong. We spoke a little of everything. Tales of war, scenes of battle, sunsets seen yonder, in the Straits—unique sunsets to which even the rough lads who had fought there had not remained insensible. We spoke too of the successive bombardments undertaken by the Allied squadron. And this is where my story begins.

The day before, a bluejacket had come to our ward. He had promised to return and bring his share of anecdotes. And now, as it happened, here he was! He told us the following:

"When, on the 19th of February, we received Vice-Admiral Carden's order to go and reduce the fort at Kum Kale... how glad we were! I believe we would have worked night and day to hasten the moment that would allow us to take our turn in the fight... Oh! how glad we were."

Here, he breathed deeply, expanding his chest... evidently the memory of that memorable day was alive within him...

"Ah! Our *Suffren*![1] how we had tended her. We set off at full speed. And scarcely had we reached our station when we began indirect fire at long range (11,000 metres), then we advanced to 6,000 metres... Our fire was well-regulated and we quickly did good work. So, when the British battleship *Vengeance* came up, flying the flag of Rear-Admiral de Robeck, intending to make a dash at close range, it was Kum Kale alone that did not answer. The three other works—Helles, Sedd-el-Bahr, and Orhanie—did not hold back. The poor *Vengeance*, thus bracketed, was a prime target and it was impossible for her to return fire on all sides. They were firing at her from everywhere. It was an unbroken cannonade! What a din... Our own *Suffren* was shaken by all the external concussions; the whole ship shuddered."

1 Commissioned in 1902, the 12,432-tonne *Suffren* was a pre-dreadnought battleship armed with four 305-mm (12 inch) and ten 164-mm (6.5 inch) guns. The ship was assigned to the *Escadre de la Méditerranée* (Mediterranean Squadron) for most of her career and often served as a flagship. On 26 November 1916, she was torpedoed by the German submarine *U52* off Lisbon; the magazine detonated, sinking the ship instantly with the loss of her entire crew.

"So the *Suffren*, seeing the danger the *Vengeance* was in, immediately came about to port to bring the Helles battery into her arc of fire. Now it was our turn to open up. Our ranging took no time— and precise it was... Then, without losing a moment, we sent over three magnificent salvos that kept her quiet... The British said we had 'silenced' her. They marvelled at the alacrity with which we had sprung into action. They never tired of telling us so, and ever since, throughout the British fleet, the *Suffren* has been nicknamed the 'fire-eater'."

And he went on:

"One day, I shall explain to you what a warship really is."

———

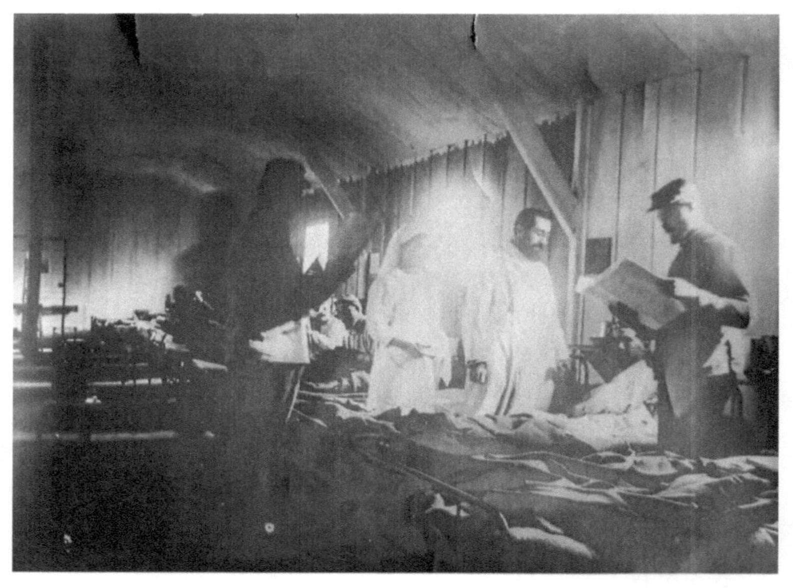

Interior of a ward at Evacuation Hospital no. 1, Mudros.

(Elisabeth Jardin Collection)

III

Mudros, September 1915.

When the nights are truly beautiful, and the heavens hold that great cleansing wash which scrubs the stars—making them appear larger, brighter, perhaps even smiling, in the immense azure frame they delight to inhabit—we go, as to a procession, to see them a little closer...

It has become the hour of rest, the hour to which one has a right after rude days of labour... One need only climb up there, atop any hill... One ascends with a brisk step, for on these very clear nights there is a lightness that lifts one up. You scramble, you climb high, ever higher, then, in the hollow of some rock still warm from the day's heat, you sit down... you breathe and you dream... Oh! how one dreams of peace, up there on the hill... how much nearer to God one feels...

Here, dominating all the stars, are the magnificent planets... Venus, Jupiter. And then the constellations: the Northern Crown, Sirius, Orion's Belt, the Pleiades... Sagittarius, too warlike... To name them, one would have to look at them one by one... And this race through the stars would threaten to take very long. To love them, it suffices to understand them; and since stars are very sensitive persons, they never remain indifferent. Some of them twinkle, making soft eyes at you with a gentle smile. Others hold your gaze to give you confidence. They mingle together while keeping their distance, and I wager they know one another.

Last evening, the night was very beautiful. From north to south, from west to east, not a single cloud. The atmosphere was limpid and the sky so wonderfully luminous that one fell silent. There was a vast blue everywhere—a blue wandering over the sea, a blue wandering over the great hill opposite, the one with the feline air. And as that hill is entirely covered in beautiful red earth, all this blue playing upon it, enfolding it, made of it a great living and sacred thing, very warm and very luminous, filling the eyes with beauty.

Above all, because in the valley that lay between us, there was silence—the great silence, the true silence... The silence of lands where no living soul dwells, the silence of lands possessing neither trees nor any vegetation whatever... One heard only the sound of the earth... Do you know it, this sound? The sound of the earth, yes!—That mysterious, unsettling sound, the sound of the earth which, with the coming of night, awakens, breathes, and vibrates...

Never, until last evening, had I heard it so—that indefinable, immense sound which composed the silence, which created the silence...

A stone I had dislodged tumbled down in fitful

bounds, making a dull, discreet, and uneasy sound. It was the hour of the earth, the hour of its awakening, for it was the hour of silence. And never before, as on that night, had I understood it so clearly. One felt, dimly, that here lay the great mystery of the infinity of worlds, and that this silence demanded respect...

Then, without truly understanding why, I gazed at the stars... They were old friends whom I did not fear, and who might, perhaps, tell me the secret of the future... And then—whether from reason or madness—I began to forget my anguish; a great dream swept over me... I lived beneath the gaze of my stars without seeing them; the rising sound of the earth passed without disturbing me, and the silence itself took hold of my mind, lulling it into a deceptive slumber...

I remained there a long time... How long, I cannot say... All I recall is that I was awakened abruptly... Before me, above the hill, hung the divine vision of a beautiful golden crescent, gliding in hushed silence...

It was the moon rising, mocking me a little... Her light reached down to me... And she glided, on she glided, quite at home in a sky that spoke no word...

There are other nights, too, equally calm, equally pure, yet nights when the owls cry out their misery... Oh! how they wound, how well they know how to echo your own sorrow... And I know of nothing sadder than this solitary song, this hideous, sepulchral laugh. On such evenings, it is best not to roam abroad, for one returns with a soul shrouded in mourning; and instead of the deep, deep rest one sought, there is, as it were, a vast rent within the heart...

Then one returns in haste, stumbling over the stones. And to fortify the heart, if the moon is at its height, we stop in the hollow of the valley where a slow, gentle stream glides by... There, where the water ripples freely, we find the great Eastern moon, seemingly asleep in the pool... One could almost grasp it with one's hand, so close does it appear. But the water that watches, the water that sings all around, keeps good guard... We gaze because we are listening, and the beautiful murmur gliding between the stones sings... sings to the heart... And that is why, during these beautiful Eastern nights, the soul feels lighter, and we feel ourselves the better for it...

———

IV

Mudros, September 1915.

Ah, yes, he knew plenty of stories, our bluejacket, and true ones at that. And how he would draw himself up, the lad, when he said:

"You really ought to have been there for that famous day, the 25th of February, when we had to put to sea for a decisive action against the Entrance forts. Unfortunately, we had been forced to interrupt our fire for six days because of bad weather... Well...

"The Allied squadron was made up like this: us, with the *Charlemagne*, and the British section had HMS *Vengeance* and *Cornwallis*.

"We had left Tenedos with light hearts; at last, we were going to fight. Well, the firing started at 10:45. The British ships dashed forward first. Us, we waited... It was a jolly fine sight to see those ships racing forward...

"The noise of the broadsides made a hell of a din and we could no longer hear ourselves speak. We, who were held in reserve off Cape Tekke, watched this duel to the death from the spar-deck and the side-decks. We had but one thought: to take our turn. We were jumping with impatience. Especially when we saw the mighty battle engaging between the dreadnought *Queen Elizabeth* and the Helles fort... Ashore, the 38-centimetre shells kicked up immense columns of dust and smoke—everything vanished... But every two minutes, two great flashes spurted from the cliff, and then it was the dreadnought's turn to see two huge sheaves of water rising right alongside... They towered over the huge battleship. We just kept watching, and I promise you we would gladly have been in their place.

"Helles answers, won't shut up; the fire redoubles, gets faster, the frightful howling of the 15-inch guns quickens. It's an impossible racket, while ashore the shells circle closer to the battery position...

"Can you picture it?...

"The Turkish guns were spotted; then, one after another, you see three mountains of earth rising up. Everyone cheers, the British ships cheer, everyone is wild with joy. And you can understand that, you know. We were all—all of us—ready to sacrifice everything to ensure victory...

"Soon, we see the signal flags run up the halyards of the battlecruiser *Inflexible*—that hero of the Falklands flying Vice-Admiral Carden's flag—and the *Vengeance* and the *Cornwallis* dash forward to storm the forts... It was magnificent... We couldn't keep still... They took our very souls with them. We weren't on the *Suffren* anymore, but on the British ships...

"But luckily for us, our turn was coming—action stations sounded—everyone ran to their posts... The next day, the Lieutenant told us about it... He was in his rangefinder tower, he could follow the 'run' of the two British battleships...

"They don't hang about, firing from both sides— what we call rapid fire... The racket is even more terrible... The Lyddite spreads out in a yellow cloud all around them, yet you can't make out a single fall of shot nearby... We thought the Turks had had enough, and that there'd be nothing left for us.

"When we see the *Vengeance* coming up on our port side... we were scared, us lot, that we'd be left with our arms folded.

"But then the order came to proceed: twelve knots... At last, it was our turn to join the battle. We shaped a course for the entrance to the Straits... Inside the conning tower—you know, that sort of turret from which the ship is commanded—the call went out:

"— Range to Orhanie?...

"Two angles on the sextant, a shot from the rangefinder, a glance at the pre-calculated firing table.

"— 8,500! 8,300! 8,000...

"Then, all at once, what a din—a blast of hot air whipping against the body, smoke everywhere... Our beautiful little forward 30-centimetre turret had just opened fire on Orhanie.

"We heard again:

"— Angles? Range...

"— 7,400.

"Then we let fly—it made a devil of a din! All five of our 16-centimetre broadside guns were spitting fire as if they had done nothing else their entire lives... They did not stop; they went on and on. They fired ceaselessly. What a din, my God, but how proud we were, if only you knew...

"Suddenly, a bell rang out from the voice-pipe, and we heard:

"— Shift target...

"No more gun fire, nothing, absolutely nothing. What a contrast with the moments before. Never had I known a silence like it. We were struck by it! Just think of the din that had preceded it. We heard someone say:

"—That's him done for. Whose turn next?

"The conning tower spoke again...

"— Range to Kum Kale?

"— 3,100.

"Happily, we were not done yet. The din started up again. We fired anew, striking harder and harder. And amidst all that noise, you could hear the commands being bawled out:

"—Ram home! Stand by! Fire!'

"Our *Suffren* kept forging ahead; she wasn't afraid, I assure you. She came right up to the first minefields, then came about to port. We began firing again.

We opened up on Sedd-el-Bahr at less than 2,000 metres! The walls of the castle were crumbling; it was a treat to see. The stones came tumbling down, dragging great chunks of the fort with them. Things were cracking, breaking apart. We could see their gun barrels flying into the air—and those were big naval guns, mind you. It was a 'slugging match' like no other... We kept on firing. They answered less and less; our fire struck true, destroying everything that needed destroying. We had some crack gunners on board... Soon, they answered no more... All was silence over there... The victory was ours..."

He promised to return...

———

A view from the summit of the hill above
Evacuation Hospital no. 1.
(Elisabeth Jardin Collection)

V

Mudros, September 1915.

Up there, one generally felt a great sense of well-being, a sort of release.

Up there lay the summit of the hill overlooking the camp. It took but a few minutes to reach. The air was bracing, one breathed freely, and then, above all, the view was marvellous.

Opposite lay the roadstead, with its inlets, its usual complement of vessels of all kinds—its skiffs and yawls, its battleships and steamers; to the left, a succession of camps, and the signal station all a-flutter with flags; to the right, yet more camps, and a vista opening onto the valley where a lovely village lay slumbering... And right behind, another valley—very deep, very rocky—sheltering in its folds a sheep pen built entirely of stone, held by no mortar, which two fig trees adorned magnificently...

All this was yours when you went up there! We used to climb up around six o'clock, along steep paths where the pebbles rolled and crumbled away. Twilight was just beginning to emerge from the mists... We would sit on a few large stones...

To-day, the weather was perfectly clear; a light all blue drifted through the air, tinged with mauve. It was very soft. One could trace the slightest details, that exquisite finish which belongs only to the light of the East!

Then came the twilight with its thousand superimposed hues. Everyone fell silent... Only the eyes lived an intense life... There was a kind of ecstasy in them, widening to capture it all... And in the depths of the pupils, there was a sense of wonder...

And then it was night, slipping in a few shadows with difficulty, almost with regret... Then voices murmured what they had not wished to say until then. We spoke of the East... We shared memories in hushed tones... We spoke of the war... We fell silent again... Then, someone dared...

"It was," he said, "last April... We were anchored off Skyros: a man still very young, an officer, bearing an illustrious name, a poet of merit, of Scottish origin, had gone ashore the day before to dream upon the isle.[1] Wandering with a friend—a musician just as he was a poet, inseparable companions both— they had roamed at length amidst the countless fragments of white marble blocks, scattered profusely across the island; blocks that rose from the earth, formed lines, and then were lost in the

---

1 Rupert Brooke (1887–1915), the celebrated poet, who died of sepsis off the island of Skyros on 23 April 1915. Brooke was actually English (born in Rugby). The musician mentioned is his close friend William Denis Browne (1888–1915), a composer and critic who helped bury Brooke on Skyros; Browne was killed in action at Gallipoli two months later.

distance. A few sparse olive trees were visible at intervals. Moved by a common impulse, they had stretched out beneath one of them. Both, no doubt, dreamt long dreams, for the hours passed and they gave no thought to leaving.

"Twilight caught them unawares, followed by the night. So they took the path back to the ship.

"That very evening, the poet felt a deadly fever shaking his entire frame. Had the island—jealous of the secrets he had come to surprise—cast some spell upon him? Dreams haunted his mind—other dreams, glimpsed through the delirium of his words. He was heard singing of the death that hovered over him. He spoke of the marbles that had doubtless served the Spirits of ancient Greece... He recited, one by one, the verses he had composed, with their rhythmic cadence, their divine flight... He struggled also to vanquish the sickness, in a final revolt, a last surge of his youthful energy. Then, with the morning, he began to feel the chill of the tomb creeping into his limbs, paralysing his mind... And then... at the hour of twilight, he died, wearing still his poet's smile.

"The story I tell is no made-up tale... Listen to the end!

"The order had come to weigh anchor in the night and sail for other, unknown lands. It was decided then... Nay, listen...

"A flag served as shroud... Then, despite the night—amidst the night—the coffin was lowered into one of the boats waiting at the gangway... Four torchbearers flanked it. And gently, in silence, oars scarcely grazing the very blue sea, we glided towards the island which the white marbles adorned with a magnificent radiance...

"It was a luminous night that evening, so pure and crystalline that every star was embellished by it... The moon radiated in the immense setting— beautiful, perhaps too beautiful—cold and haughty.

"The boats moved off very gently in close, uneven ranks... And then, because of rocks that could not be seen, we were forced to halt...

"Then... Listen on... Listen to what follows... This is no tale, I have told you... This happened last April... Men entered the water. With arms outstretched, they raised the dead man above the waves that gently clashed together...

"There was, coming from the open, that indefinable sound—gripping in its mystery; it was the sound of the waters we disturbed, the inner sound rumbling within oneself from emotions heightened tenfold... It was the sound of the fire gnawing and tearing at the resin coating the torches, the sound of great, wild flames rushing forth, losing themselves in the night... riding over the coffin, over the men, over the water... It was all that sound—contained and immense—that composed the silence... for silence there was..."

"Then, across the deserted island, amidst all those marbles lying pell-mell, came the rhythmic tread of the procession, preceded by bagpipes. The fifes sounded the march, leading the dead man towards the solitary olive tree, where he had dreamed for so long… so long… until death… The bagpipes played… played the old airs of the lochs and the moors…

"And while men, by the light of the torches, dug—right beside the coffin—the grave that was to receive it, the bagpipes played… played the old airs… those of the lochs and the moors…

"Only when the coffin slid into the earth did the bugles sound… They sounded long… The torches drew closer, meeting above the gaping hole as if in a final farewell to the dead…

"The lifeless leaves of the olive tree flushed red in their glow, seeming to embrace…

"And the bagpipes sang one last time, bidding the dead poet the farewell of the lochs and the moors, the farewell of that Scotland, misty and luminous, the farewell of the homeland… The bagpipes sang the farewell of the living… The bagpipes sang long… so long…

"After the bagpipes came the slow bells of a flock lost among the rocks… A shepherd's pipe sang out in turn in the distance—some herdsman expressing his own thoughts in his own way!"

The storyteller fell silent… None of his listeners thought to interrupt the silence rising all around them. Only, they held within their eyes, within their hearts, within every fibre of their being, the island of white marble, the nocturnal procession, the sound of the bagpipes weeping and laughing all at once... the solitary olive tree... and the torn earth... And the little bells, too, that went *dring... dring... dring... drong... drong... cling... cling... cling...* in the night... And the shepherd's pipe...

A sound from the sky, cold and sinister as a guillotine blade, passed over our pensive heads... Startled, everyone looked up... An immense flock of ravens was making from south to north. And the black wings of the birds snapped... snapped... whipping the sky in their headlong flight...

———

VI

Mudros, September 1915.

Our bluejacket, who had left us for a time on duty, has finally returned. This time, his eyes shine even brighter than usual. He has laid in a store of memories. All the patients wait anxiously; for them, he is a true ray of sunshine.

And so, when he begins, not a soul breathes a word.

"You know," he says, "I have not told you everything. I kept the best for last. To-day, I am going to tell you in detail of that famous attack of the 18th of March against the five great forts of the Dardanelles.

"That day, I assure you, it was worth being on the *Suffren*, worth being a sailor. Just think—we were bringing sixteen battleships into action, along with all the flotillas of destroyers and minesweepers— and all for a direct, all-out attack on the Narrows.

"The chiefs had studied the problem. You know the lie of the land in the Straits. The Asian and European coasts can easily concentrate their fire on the attacker, whilst we could only bring a limited number of ships into the line to engage the enemy works at effective range. Yet the channel had to be cleared; without that, nothing could be done. We had to stop the harm being done to our sweepers by the batteries situated at the water's edge, abreast of the great minefield.

"It was no small business, as you may well suppose. Think of the magnificent target we offered the enemy. To right and left there were guns, and they were great, solid forts that we were going to attack... The official reports have stated in their time what scheme was settled upon... But I—I am going to explain it to you..."

Here, our lad pauses as if to marshal his memories, to take them up one by one so that everyone might understand—and above all, so that there be no mistake.

"Well then, they had decided to send a first line of four British battleships to lie across the Straits, 13,000 metres from the Chanak–Kilid Bahr line. Their mission was to bombard the five principal works with deliberate fire, whilst naturally remaining out of range...

"You can picture it... They had to wait for the proper moment—that is, when we were assured the forts were sufficiently knocked about to allow a second line of four older battleships to advance 4,000 metres ahead of them. These latter ships were ordered not to foul the fire of the first line, whilst attacking the same great works just as they did, and engaging the secondary forts within reach. It was no easy task, that mission... But it was ruddy fine work... Between the first and second lines, a flanking battleship was sent along each shore. For

I must tell you, the field batteries did not hesitate to spit fire at us, and they had to be silenced... Everything was provided for, and a relief force was to replace the four battleships of the forward line and the two flank-guards about two o'clock."

"Well, that vanguard post—the four old battleships—fell to us French. And we were not a little proud of it. Just think, it was the most dangerous post... We had to manoeuvre carefully so as not to mask the fire of the four modern battleships. Our division therefore split into two sections—one to operate inshore under the Gallipoli Peninsula, the other—us—along the Asiatic coast. You know that beautiful coast that stretches out, where the light is so lovely. But that day, it was especially dangerous; so our Admiral,[1] who feared nothing and knew a thing or two about courage, claimed it for his own section: *Suffren–Bouvet*.

"Now that we were in the thick of it, we realised the difficulties. We were right by the mines; the water inshore was shallow; and as we were obliged to keep our target on the beam to engage it with our full broadside, it was no easy matter. Think of the narrow margin between the ordinary limit of our guns—the medium calibres—and the initial firing range which the current increased so rapidly. All this made for a set of circumstances that cramped

1 Rear-Admiral Émile Guépratte (1856–1939), commander of the French squadron, renowned for his gallantry and aggressive tactics. He flew his flag on the *Suffren*.

our movements more and more, so that our field of action was reduced to an almost mathematical point. The ship charged with firing on the great forts had to stop, even though she drifted... The second battleship, keeping station about 500 metres downstream, had to fire on the secondary batteries whilst standing by to come to the other ship's aid and relieve her..."

"Besides, the chiefs—as agreed beforehand—had decided on changes of station, as much to distribute the strain on our guns as to allow us to counter the current—without having to interrupt our fire for all that..."

"You see, the *Suffren* and the *Bouvet* were the target for the three great forts on the European shore—Yeni Medjidieh, Namazieh, and Rumeli Hamidieh. The three greatest forts!... And what is more, we knew full well that we could successfully engage only one of the two great Asian forts—Chanak and Hamidieh. As for the batteries at Soandere and Dardanos, we knew perfectly well they would not be silenced so easily, for they were devilishly well-armed, and above all, supported by field guns..."

"But that mattered not a jot; we were resolved to fight. We knew the business would be of the hottest, that all manner of difficulties would arise—so be it... we had to go in, and we went in with a light heart, never forgetting what we were risking. One could well sacrifice one's life for one's country. And a death such as that was not a death to be scorned...

"Now that I have fully explained our positions, and you know the Dardanelles—you can see the Asiatic shore and the European shore?...

"Well then... listen.

"The modern battleships, obeying orders, had opened deliberate fire at 1100 hours. The great guns pounded heavily, pausing as if to listen to their own roar, then resumed. The air held nothing but the sound of the guns; it was the cannonade that blotted out all other impressions. Rear-Admiral de Robeck—who had relieved Vice-Admiral Carden in command of the Allied fleet the day before—had ordered us at 1215 hours to proceed to our station. We split formation immediately. The *Gaulois* and the *Charlemagne* made for the European coast, whilst our *Suffren* and the *Bouvet* made for their designated point, steaming at 12 knots towards the Asiatic shore. We arrived within 9,000 metres of the Kilid Bahr forts... And at precisely 1240 hours, our ship opened fire on Yeni Medjidieh...

"My friends, scarcely were we in position when a rain of shells fell all around us, and around the *Bouvet* too. They came in every calibre. Great plumes of water rose up; there were battering-ram blows against the ship's armour, as if enormous, prodigious hammers sought to stave it in. It seemed the whole ship must give way. Add to that the internal shuddering of the engines, the recoil of our own guns, the intense, undreamt-of vibration that shook her to the core.

"Despite this, we carried out our programme to the letter. Our two ships obeyed orders punctually. They relieved one another as agreed, so that the main objective did not cease to be under fire for a single minute... Each of the two ships occupied the firing station twice, for twenty minutes at a time...

"For our part, we now existed solely on sheer nerve and heart. We no longer thought of ourselves; we thought only of the guns, of our ship, and we were proud. Our *Suffren*, after her first 'round'—during which she had closed to the very limit of firing range—had sustained only two minor hits, though God knows the enemy had favoured us generously enough with their shells. But the *Bouvet*, taking our place, was soon ablaze with two fires, her poor forward turret put out of action. Yet this did not stop her from carrying on with her work as if nothing were amiss.

"Our *Suffren* returned and resumed her station— upon which, as you may well imagine, the enemy had not failed to find the range. In less than a quarter of an hour, we took a dozen heavy hits, one of which glanced into Casemate 10 and Turret 6. We lost twelve men there... Great jets of flame and smoke were seen in the port magazines, and fires broke out in the boiler rooms and between-decks. Worst of all, the fire-control circuit for the port battery—the side engaged—was put completely out of action.

"But it was not over. We had not ceased firing

when a leak was reported forward on the port side. We concluded that the port bunkers must be flooded, and the ship began to list slightly... The forward funnel was almost demolished at its base... We could still hear the din, we could feel the ship listing, but we were not afraid.

"Poor *Bouvet*, despite the blows she had received, came immediately to our aid, enabling us to open the range and bring our starboard side to bear on the target. She continued the attack without the slightest faltering. Towards 1345 hours, we were preparing to relieve her once more, when Admiral de Robeck, realizing the intense fire to which the French division was subjected, signalled us to withdraw. Our station was to be taken by the British relief battleships which were just then arriving in the Straits...

"Our *Suffren* and the *Bouvet* had taken some punishment, that was undeniable. We had dead, and wounded... But the important thing, above all, was that we had executed our mission well. We had faced, without wavering, the concentrated fire not only of the five great forts of the Narrows, but also that of the devilishly well-armed batteries at Dardanos, Soandere, Kephez, and the Quarantine Station.[1] And then, too, we had the field guns. All told, it amounted to a concentration of fully twenty-five heavy calibre guns in action—

---

1 Several batteries were emplaced near the quarantine station south of Dardanos.

24-centimetre and 35-centimetre—and nearly as many of 15-centimetre. And I do not believe I exaggerate in putting the number of projectiles that fell around the pair of us, in a little over an hour, at four hundred."

"There was even a 15-cm shell that sliced diagonally through the bridge. It lodged itself in the chart house. It passed close enough to graze the Admiral and the Captain, for our Admiral and our Captain had stepped out of the conning tower to better gauge the state of the engagement.

"They had to attend to everything: follow the gunnery, order the necessary measures to repair damage, and watch over the handling of the ship with jealous care. We were right alongside the shoals and on the verge of the fixed minefield... On the water, we could see floats of every shape that had to be avoided at all costs... And we knew, too, that drifting mines might be released against us. You can imagine the vigilance required!

"But we had not laboured in vain. All the great forts had fallen almost silent.

"But," said our big lad, "that is not the end of it. Our magnificent *Bouvet*—she whom we called our valiant consort astern—met with a worse fate than ours... We were returning; it was exactly 1358 hours, and she was about 500 metres astern of us, when we saw her list suddenly to starboard... It happened with incredible speed. She listed to nearly 50 degrees. We could not understand it... We saw a little smoke that seemed to issue from the starboard 27-centimetre turret, but we heard not the slightest explosion; there was no plume of water, no debris... We were still watching, uncomprehending, when after a pause of twelve or fifteen seconds during which the stern settled and the list seemed to check, the poor *Bouvet* suddenly capsized... Her keel stood out against the deep blue sea, and then she disappeared by the stern. It happened so fast... On her bottom, all green with algae and weed, we saw men running... Almost immediately they were cast into the sea, engulfed... The disappearance of the *Bouvet* took less time than it takes me to tell you of it... Less than a minute... It was devastatingly sudden...

"Every possible launch was lowered, from the British side as well as the French, regardless of the Turk's continuous fire... But so few men were saved..."

———

PART III

Mass celebrated in a *marabout* overlooking
the harbour and town of Mudros.

The officer standing at centre is Dr Léon Duchêne-Marullaz,
chief medical officer of Evacuation Hospital no. 1.

(Elisabeth Jardin Collection)

I

Mudros, October 1915.

As I have not yet spoken to you of our Sunday Masses, I shall do so to-day, if you will permit me!

Ah, yes! How simple they were, those Masses which, beginning at five o'clock, continued until late in the morning. In our unit alone, there were fully thirty priest-orderlies,[1] not counting those who came from outside. And as religious duties had to be fitted in with the schedule, Masses were said constantly—in the small *marabout*[2] that served as a chapel, and in the little stone shack where a great vine concealed the door. It was a Greek chapel that seemed abandoned, although a few fresh graves lay at its feet... One bore the name of a certain Catherine, aged twenty, who had died in 1914; the white inscription on the great black cross could be read from afar... A dozen or so graves were ranged there against the barracks that housed our sick...

To return to our Masses, I shall tell you of two of them... Yes, there were Masses at every hour of the morning, but to make quite sure, I had arranged

1 *Prêtres infirmiers*—ordained priests and seminarians who served as stretcher-bearers and orderlies in the French army medical service.

2 Bell tent.

matters with one of my orderlies, and he waited only upon my pleasure.

Sometimes at eight o'clock, sometimes a little later, depending on when we were free, we set off, he and I, escorted by another priest, and made our way up the hill towards the tiny *marabout*. There he put on his alb, vested himself, and began his Mass, which he said slowly, articulating the words clearly.

Further off, the same ceremonial was taking place for others. Yet no one remarked upon it, and all remained free to do as they liked... I remember a day of rain, of bitter wind... Impossible to stay outside. The flag covering the high altar was flapping wildly. We had to bring the altar inside—and ourselves with it. For try attending a Mass in rain that lashes and pierces you. So I found myself behind the altar, sitting on the reverend father's camp-bed, to leave more room for those who were abandoning the other Mass to come to ours. For my orderly, too, had a reputation for holiness that won all hearts.

It so happened that among the congregation were several who knew the hymns usually sung in praise of the Lord. They did not hesitate to sing at the top of their voices, so that our whole *marabout* was filled with very melodious singing. And I assure you that under that turmoil of dust, rain, and wind, it was still we who were the stronger. The voices rose above the storm, and then, when they ceased, one heard the calm, pure voice of the priest who, not wishing to lose a single word of his office, spoke in a very pure, very holy tone. Then came the responses, and we who faced him across the altar, watching him, felt ourselves filled with a deeper reverence. Ah, yes, those Masses said up there were good Masses. The priest and we were but one, packed together as we were. Yet it was so simply natural... It was so open. The good Lord, I think, was never closer to His flock.

Another Sunday, we took the tiny Orthodox chapel by storm. As it happened, a priest had just finished his Mass. We quickly took his place.

The altar was hollowed out of the stone, and there was barely room to set down the ciborium... One could scarcely fit six people into that hovel. But it had its charm, an ancient, wholesome air, a primitive look that gave it dignity. On the wall hung a few icons, while the vine covering the door let a large leaf, all russet with autumn, hang down between two stones... As for seats, there were none to be found... But in a corner, a battered crate had been tossed. To take it, shake off the dust, then drag it to the middle of the room between the two arches and sit upon it, was the work of a moment. The earth served as our *prie-Dieu*,[1] and while the rain continued to lash outside, we, safely sheltered, gave thanks to God.

1 Kneeler.

Then, the office ended, in that doll's house setting where every word stood out clearly, we set off again together just as we had come. The priest became a soldier once more, and life resumed! Ah, yes! Those simple lives, that life where one can be oneself, that life which leaves only the best of you. That life, you see, is the only true life. One must enjoy the life God grants you, one must enjoy the lives of others, because, you see, life must be earned, and to earn it, one must understand it...

———

(Elisabeth Jardin Collection)

II

Mudros, October 1915.

Our young bluejacket has come back to us. He is
as talkative as ever, full of spirit, and he loves his
trade so much that he needs no coaxing to tell us
what we know so little of. Besides, for my patients,
these tales of the sea are something quite new, of
which they know scarce a word. They have never
visited warships. Those who have seen them have
seen them only from afar. And so, when our sailor
explains:

"You know, warships aren't like merchant
ships. Our machinery, to be protected, is further
underwater; we draw more depth. Our engines
are more compact, they lack the spread, the velvet
smoothness of the great luxury liners. If you have
never seen the engine rooms, you cannot know. It
is an uninterrupted tangle of piping, inextricable
at first sight. Small and large, they run, return,
double back... Great red guts that look all covered
in blood. All this is packed tight, held in, so as not
to take up too much room. Thus the transmitting
station, to the uninitiated, appears like a diabolical
lair. Impossible for a layman to find his bearings.
It is nothing but electric cables, switches, dials,
rangefinders, voicepipes, etc.

"And in the turrets, you have the handling of our great guns which glide as if they weighed nothing. All that works electrically. You load, you unload, it works marvellously. A warship is a great city, where one lives much more under the water than above it...

"When action stations has sounded, there is no one left on deck. It is like a ship dead on the outside... But what life within! The ear-splitting Klaxons, the command signals, the swarming in the bunkers of those who tend the boilers.

"Everyone is at his post and each knows what it will cost him if an accident happens. Impossible to flee. You are sealed in, trapped... Not long ago, an accident happened aboard a British ship. It was in the torpedo flat. They were forced to flood it. Fourteen men were inside; none could escape. The door was shut!"

We shuddered...

He went on:

"But the same thing happened to us aboard the *Suffren*... During the attack of 18 March, in the Dardanelles... A heavy shell fell into a casemate. We had twelve little lads there... All twelve remained. We found their bodies shredded, charred, unrecognisable, right up against the guns they had served to the very end..."

Then he continued, his eyes fixed on the distance as if to recall it better...

"We went out to commit them to the deep off Tenedos. That morning was cold and mournful. The sea itself resembled a shroud. I recall that morning as if I were still there, as if it were still before my eyes. The crew had been mustered on the quarterdeck, in their Sunday rig. We were all lined up along the rails. And not a word passed between us. Yet we knew that death was almost a friend... to a sailor!... Our hearts were heavy... Why?...

"At last, from a hatchway, our bugle sounded:

"—Attention!

"The Captain arrived with all his officers, escorting our Admiral.

"He spoke to us all, our Captain. He told us of the pride he felt in commanding men like us, and that he counted on us to uphold the honour of the flag and carry it to victory. He spoke of the dead; he told of the deep emotion, the sorrow he felt at parting with these children who had always done their duty.

"At that moment, the wind which had risen snapped our half-masted colours at the mizzen gaff... And you know, the colours, when one is on the high seas, are something more than our blood, than our soul—they are more than anything... Our colours—how ready we were to give more than our lives to keep them ever proud and defiant...

"The sky refused to clear; a greyness still hung in the air—weather truly fit for mourning. The sea, however, was not rough; it merely lapped, huddled in on itself as if it were cold...

"The Admiral and the Captain, along with our officers, then went to the armoury. They wished to pay their respects to our comrades. They had been laid out side by side, each body wrapped in grey canvas... The absolution was pronounced... The band played a funeral march... As for us, we were seething up on deck; how we longed to put to sea to go and avenge them...

"Afterwards, as is the custom, we bore them one by one to the gangway... A plank had been rigged there as a slide... Our chaplain followed... He blessed the sea... The quartermaster piped the honours. Then we heard three volleys of rifle fire, followed by a dull, flat sound—a harsh noise that gripped our hearts. We counted twelve. Each time, the water answered with a splash, sending up spray that struck the hull of our ship. Air bubbles rose to the surface; that was all that returned to us of our dead. Never had the sea been so mournful...

"The most heart-rending sound I heard was the sob of one of our quartermasters. Ah! God, how he had fought to hold back that sob... We dared not look at him, so greatly did we respect his grief. And then, in the end, it proved the stronger. He stood rigid at his station, but when he saw his young brother disappear—well... we heard a sob. And his eyes—

they were like the eyes of a mother whose child is being taken from her... It was no cheerful sight, I assure you...

"Even the British signalmen, who had been amongst us for several months, were not unmoved. One of them—a man with large, soft blue eyes and a girlish complexion that was strangely at odds with his powerful build—had tears beading on his lashes...

"Then, they played the *Marseillaise*..."

———

Nurse Elisabeth Jardin in gumboots on
the plain before Mudros town.

(Elisabeth Jardin Collection)

III

Mudros, October 1915.

Last night, a tremendous downpour from a violent storm woke us with a start. It was raining inside our hut! It rained on our little iron bedstead, which threatens to collapse each time one turns over. It streamed down the partitions, and thanks to the furious wind blowing from the south, I received on my head and shoulders a proper soaking.

For several days, I had been running a high fever. And all this cold that penetrated my hut chilled me to the bone. Water ran across our floorboards. We had to place stones under our trunks to save them a little. We spread waterproof canvas everywhere and slipped underneath it. We opened our umbrellas and waited for the rain to decide to stop... This foul weather lasted for three days...

In their huts, our poor patients had been soaked through. And despite the pitch-black night, we had to move them elsewhere. We crowded them into cramped rooms where the rain was less heavy...

You should have seen the wretched state of our hospital when the fine weather returned. The mud came up to your ankles, and in places, it was higher than our knees. A greasy, clinging mud into which one sank and which held you fast. We had to use walking sticks for support if we didn't want to go sprawling full length...

During those rainy nights, when the sky is so black that you can see nothing, and the wind is blowing and your lantern goes out, I assure you that one needs a sharp instinct to find one's way. You stumble at every step, you slip, you pick yourself up, you trip over wires, and then... you start all over again...

As soon as the sun reappeared—and when the sun returns, it's a grand occasion in the East—we put out the poor straw mattresses, yawning now more than ever. We spread out the blankets, which weighed heavy with all that water in their wool. The greatcoats were spread wide, and we roasted ourselves a little too, simply to feel a bit of life running through our veins...

———

IV

Mudros, October 1915.

The foul weather brought with it a fresh outbreak
of insects. Two enormous tarantulas, which were
wandering about, stretching their long hairy legs
inside our mosquito nets, fell victim to a sustained
and victorious hunt. An enormous centipede was
found in the sheets of one of my companions. He
was cold, the poor beast—we killed him too!

But it is the rats on the roof who kick up a devil of
a row each evening. Obstacle races, trotting races—
we have the full gamut. The mice, however, hold
the record. Entering my room, for which they have
a marked predilection, it is a delight to see them
running themselves breathless. Then, the moment
I blow out my candle, they go at it with a will. They
tumble onto my face, run across my bed, climb up,
climb down, all in truly excessive haste. Sometimes
a frightful crash wakes me with a start. It is my mug
that they have dragged behind them, rolling across
the floor.

Alas! They are not always very proper, these mice
of the Orient, and when day breaks, I note the
evidence not without wrath.

I must say, to my shame, that I do not bear them
all the animosity I ought! And although they lack

the grace, the velvet softness, and the wit of their tropical sisters, I cannot help but admire their suppleness, their tininess, and those shrewd little eyes shining from their brown coats... Naturally, I hear your cries of horror, but you see, when it comes to beasts—whatever the kind—it is of them that one need have the least fear...

I should have liked to take you into our pantry, at the hour when our hut goes dark. By the light of a candle, arriving by surprise, you would witness the maddest of spectacles. It is no longer just 'mice', but hundreds of mice gliding from all sides, hurling themselves from top to bottom, jostling and trampling one another, darting forward then falling back, vanishing into the lined-up shoes, slipping from vases, emerging from cups, tumbling down the partitions, losing themselves again in a mad dash, during which, imagining that we are out for their blood, they flee... they flee...

———

V

Sedd-el-Bahr, October 1915.

We took a little over twenty-four hours to go from Mudros to Sedd-el-Bahr! And that is a journey one usually makes in six. Mind you, the wind and rain had shaken the sea so thoroughly that she was in quite a temper… And then…

Well, I shall tell you the story of our voyage. We had weighed anchor around five o'clock in the afternoon, when, at eleven, I was woken with a start by an unusual coming and going in the corridor. Voices could be heard: "Fire on board." "Where?" "In the hold." As we were a hospital ship and carried no munitions, it was hardly worth disturbing oneself for so little. I turned to the wall and went back to sleep.

At two o'clock, another awakening. This time I heard no noise, but the motion of our boat was unmistakable. I had no doubt. We were drifting, for lack of steering. The ship was rocking gently on the water, and for one accustomed to the sea and brought up on its tales, one knew very well what that meant.

A few moments later, I heard hurried footsteps again; I could sense the anxiety they were trying to conceal.

I looked at the time. It was indeed two o'clock! The night was black. Without a doubt, it was cold on deck. I was warm. And, all things considered, it would have served no purpose to climb up there. So I fell asleep again—a sleep so peaceful that it was nearly seven o'clock when I awoke.

We were anchored off Kephalo...[1]

Ah! What lovely light was playing over the island. I assure you my porthole seemed much too small to satisfy my desires. I hurried up to the bridge.

There, the captain told me the story. The fire had been contained. They had extinguished it, but it had had time to reach the steering gear. The boilers, owing to bad coal bought at the Piraeus, had been more than sluggish. And for a few hours, he had no longer been master of his ship. The current was carrying her straight as a bullet towards the Asiatic coast, which we had approached to within barely two thousand metres. Fortunately, another current had intervened, and, repairs helping, they had been able, little by little and not without difficulty, to work back towards a less dangerous zone. And then, too, we had passed alongside a great floating mine drifting by. Ah! The poor captain, with his clogs going *click-clack* on the planks—he was only just

1 Cape Kephalo, Imbros island, some twenty kilometres west of the Gallipoli Peninsula. It shelters the eponymous bay that became an important staging post for operations in the latter half of 1915. Kephalo was also the location of General Headquarters, Mediterranean Expeditionary Force.

beginning to regain his composure! What a fine man he was. He had seen it rough at sea. Before becoming a hospital ship, this vessel had served as a transport. And you should have heard him recount that landing at Kum Kale, all those ships waiting off Sedd-el-Bahr while they made a diversion on the Asiatic coast to allow the troops to land on the European shore.

"It was black with ships... the guns thundered, the shells fell... There were gusts of shrapnel... But it worked all the same... We made it... But the men we left behind..."

———

Sedd-el-Bahr castle.

(François Masnou, Agence Rol)

VI

Sedd-el-Bahr, October 1915.

When we anchored off the tip of that famous Gallipoli Peninsula, right between Cape Tekke and Cape Helles, I felt a sudden blow to my heart. I felt cold, so very cold. There were great heavy, mournful clouds in the sky, clouds of dark omen. The glacial north wind whipped about unhurriedly, casting its graveyard shroud over the waters and the land.

I was at the entrance to the Dardanelles, and I knew what an immense graveyard this strip of land represented. Sedd-el-Bahr was there, within sight, utterly shattered, its mosque stove in. And the old *château d'Europe*,[1] half-collapsed yet still in places proudly bearing its shattered battlements... the old fortress seemed to be weeping. Its cypresses, scorched by flame, trembled as if in fear. And all along the flanks of this peninsula were multitudes of tents. The earth was beaten, arid. Further off, one could see ravaged Krithia, riddled to its very entrails, with the black line behind it marking the notorious and sinister Kereves Dere ravine—the

1 'The Castle of Europe.' The Ottoman fortress of Sedd-el-Bahr at the tip of the Gallipoli Peninsula, so named to distinguish it from the 'Castle of Asia' (Kum Kale) on the opposite shore of the Dardanelles.

Ravine of Death, as the soldiers call it. Above, Achi Baba, impregnable owing to its position.

And farther, ever farther, in the vast distance, the Bithynian Olympus[1] casting to-night—owing to the winter sky—a mass that was harmonious yet wrathful... The Asiatic coast, with Chanak dominating, quite close to us... The Plain of Troy now, to the right, with its tumulus—the Tomb of Achilles, so they say. Yeni Keui, Yeni Shehr, the Ida mountains closing the horizon, and Kum Kale opposite, quite close to us.

And then, the entrance to the Dardanelles. Rarely have I felt a sweeter impression than when gazing upon it. How beautiful that tranquil water was. What infinite charm, what calm, what quietude! Pearly reflections played upon it, reflections that stretched out, folded back, then returned and lost themselves once more.

But great flashes slashed the horizon, great flashes coming from Achi Baba. A dull, muffled sound; seconds slipped by in silence, and then, right beside us, the sharp, terrible noise that strikes and kills. White puffs rose and slowly dispersed. One... two... three... We counted, we counted again, and yet again we counted. The same flashes now sprang

---

1 *L'Olympede Brousse.* Modern name, Uludağ. Located about 255 km to the east, it is impossible to see Uludağ from Seddülbahir. The author likely mistook a cloud bank or closer peak on the Biga peninsula for the distant mountain.

from the Asiatic coast, bursting over Sedd-el-Bahr and Krithia. But life on land continued without pause, unflinchingly. Horses filed past, convoys moved, other horses neighed on the hillside where they were curiously perched. Boats came and went. The shells still fell. The slimy hull of the *Majestic* weighed gloomily upon the water, a dead shell preserving its corpses.[1] And the *River Clyde*, with its cargo-ship air, perched in full view, insolent to the end, mocking the enemy despite being riddled and rusted.[2]

From the sea, other flashes, larger perhaps because they were closer. And once again, and so many times again, seconds elapsed—seconds that one weighed, that one counted, and from which one drew conclusions. The monitors fired. They answered mechanically in heavy broadsides... And one could no longer make them out in the night that was descending, hard and implacable.

1   Torpedoed by German submarine *U21* on 27 May 1915, HMS *Majestic* capsized in shallow water with the loss of 49 lives. Her masts rested on the seabed, keeping the upturned hull visible above the surface for months until a storm finally submerged it.

2   The *River Clyde* was a collier converted into a 'Trojan Horse' landing ship for the assault on V Beach, Cape Helles, on 25 April 1915. Deliberately grounded to deploy troops directly onto the shore, she remained beached throughout the campaign, serving as a vital breakwater and landing stage.

To-day it is a radiant morning, like a holiday, with a fine, limpid light that adorns every detail with a very gentle grace. It is so luminous that the eyes are quite dazzled. Imbros lies right behind us, Imbros whose name has a pretty ring to it, Imbros with its broad, supple lines and its mauve that shimmers and fades into the marvellous blue of the waters. Tenedos, and again the incomparable Asiatic coast which draws near and fades away. Feline movements, feline grace. The immense Plain of Troy; Yeni Shehr weeping and quivering, Kum Kale disoriented, yet still keeping its line amidst the heap of ruins.

The guns thunder. The Ida mountains stand out more clearly, raising their peaks with an archaic grace. Blues and mauves still drift, lingering complacently, and the eyes wander, pause, and return. The sun lashes Sedd-el-Bahr and Cape Helles. The earth, stripped bare to its very entrails, sits passively. It is turned and turned again. Men are digging, and have dug. Men live underground, always—night and day, in the rear as at the front. Intense dust rises in moments, followed by detonations.

Nowhere else have I seen a light so pure, so beautiful, so gripping. Every hour brings with it a whole procession of new colours, unknown sweetnesses, and greater songs. And as they pass, as twilight approaches and one forebodes the final riot of colour, one shivers the longer.

Oh, the tragic grandeur, the dreadful sadness, the incomparable beauty of these evenings...

An aeroplane circled in the air, light and graceful as a swallow. And with little hops and mischievous bounds, it landed on the cliff-top, in full view of enemy fire. The sun irradiated its wings. The pilot climbed down, another ran up, and once again the bird, in less time than it takes to tell, set off, braving the bursting shells.

And I saw those field hospitals that nothing can shield from the enemy. I saw the countless graves on the shore, piled high with shingle and bathed by the sea. I saw all the other graves where twenty or thirty dead are heaped into each one. I saw that famous Morto Bay where so many of our men rest among the great black cypresses. I saw the makeshift dwellings of all those who live there. I breathed the frightful stench that exhales from everywhere. My eyes were burned and my mouth parched by the waves of dust raised by the wind. I saw all this unknown misery, all this suffering, and I knew what is endured by all those who 'work' there.

———

The 'Trojan Horse' landing ship *River Clyde* at Sedd-el-Bahr.
(*François Masnou, Agence Rol*)

VII

Sedd-el-Bahr, October 1915.

It is true, I did not tell you of one of those nights I spent anchored opposite the Gallipoli Peninsula. I forgot to tell you of it.

How beautiful that night was, but how sad! We lay upon a vast graveyard, for deep at the bottom of the sea slept so many corpses, and we were quite close also to another, even greater graveyard—that of the land.

The coast of Asia, no less than the coast of Europe, was etched in silhouette against a sky of exquisite purity. Across the whole expanse of the heavens, there was nothing but a swarming of stars, as large as they were brilliant. And they all moved off, like an immense host, with the same peaceful tread. Oh yes, how beautiful that night was. There was in the air that great mildness, that sort of warmth that holds you, that draws you closer to the sky. Up there on the bridge, one felt nearer to God.

Then the moon slipped right behind the Bithynian Olympus. We saw her climbing towards the heavens with such assurance and calm! And then it seemed always that it was of her own free will, and by condescension, that she let trail upon the earth a little of her great beam of light. It slipped over there,

The Chanak searchlight.

(Ernest Brooks, Admiralty official photographer)

on to the Asiatic coast, then came to play with the tiny waves which were beginning their dance again, the better to welcome her.

Oh, that immense sweetness! More than once, I reproached myself for having relished those hours so much. But you see, on nights like these, one understands better, because one is nearer to God. And again, on such nights, it is God who shines forth and whom one loves. Yes, one feels better, and one prays better. Over there, in the beautiful shaft of moonlight, we could see the black buoy—the one marking the spot where the *Bouvet* disappeared. And I prayed, I prayed for all her dead, as I had never done until then.

The enemy himself had fallen silent, and we, no doubt, were thinking as he did.

The moon climbed ever higher, mounting straight towards the heavens, and the stars seemed to draw aside as if to give her passage... Her great plume of light still played upon the water, startling awake those waves that were falling to sleep...

And then, this unique lighting did not suffice for those on the other side. Chanak cast its searchlight and struck us full in the face! Oh, the insult... I blush at it still... Oh, what insolence, more stinging than a lash of the whip. I understood then that even on beautiful nights in the East, the soul can be bloodthirsty.

It had caught us in its orbit and was doubtless marking us out; it held us in its talons, staring at us, scrutinising us, and then, without another word, it swept elsewhere...

———

VIII

Sedd-el-Bahr, October 1915.

Why, yes, I did visit that famous laboratory, that diabolical lair they speak of in Sedd-el-Bahr—the laboratory of Dr S—, installed in the cellars of Sedd-el-Bahr castle.

I assure you, a visit was imperative. Obviously, one had not to mind stooping to enter the single corridor that led to it. Nor should one fear the dust, or indeed the darkness. By groping one's way along the walls, one arrived almost unaided. The great ramparts that have borne the weight of centuries still withstand the shock of the shells; and just try to disturb a bacteriologist when he is bent over his glass tubes and countless phials...

But to return to our laboratory, it was set up in a domed room. There, one could easily imagine the deeds of yesteryear: the inquiries held behind closed doors, the Judgments, and then the suppressions—with neither formality nor scruple—of the offender selected to satisfy the whim of some host... All of this was quite simple, quite matter-of-fact... And just try to hear the groans of the accused through those walls built for oblivion. Heavens, never have I felt the weight of stones so heavily as on that day.

Now, our laboratory was indeed a laboratory.

Everything had found its place there, from the large microscope sleeping under the benzol lamp to the incubator that ran without a hitch. There was, to be sure, a little dust. And the light was not always quite sufficient, but by looking closely, one could still manage. The master of this diabolical lair was, when I entered, in his great white smock, carefully spreading upon a glass slide a mixture that was of a superb blue— Mytilene blue, I was told later.[1]

The lamp against which he was working lit him full in the face. And all around him there was nothing but darkness... I assure you it was no ordinary spectacle, and despite the impassivity and the antiquity of the stones of Sedd-el-Bahr castle, they too looked on in bewilderment, as if they had never conceived of such a thing...

The welcome was most charming. And while the muffled sound of shells rang out one by one above our heads, I lingered, that I might remember it well... As a souvenir, I brought away a photograph of this most unusual place—a photograph taken by magnesium flash—and I wager that, faced with such an intrusion of civilisation, the old stones themselves almost choked at the presumption...

———

1 Likely a corruption of—or pun on—Methylene blue, a standard laboratory stain. Mytilene is the capital of Lesbos, an island just south of the Dardanelles.

IX

Mudros, October 1915.

It was a queer sort of voyage, let me tell you, that I made from Sedd-el-Bahr to Mudros. The *Jeanne-Antoinette*[1] was a little over a hundred tons, and we had scarcely left the jetty alongside which the *River Clyde* was lying when a violent south wind began to blow... Heavens, it was no longer quite so amusing... Our boat, lighter than a cork on account of her lack of ballast, leapt over the waves with such gusto that one marvelled at it. The deck was crowded. A hundred sick men, with a few wounded among them, for cabins were out of the question. There was only the hold beneath us, where wine casks were usually stacked.

Although I am a good sailor, I did not find this waltz very amusing. With every heavy sea—and they were many—water came aboard. The men howled. A few sea-sick Senegalese,[2] heaving with immeasurable effort, rolled their eyes back until only the whites showed. One heard nothing but a

1 French steam trawler (built 1906) requisitioned as a *patrouilleur auxiliaire* (auxiliary patrol boat). Roughly 38 meters long and displacing just 174 tons, the small vessel served in the Aegean from 1915 until 1919.

2 *Tirailleurs Sénégalais*, colonial infantry.

Auxiliary patrol boat *Jeanne-Antoinette* at Salonika in 1918.
(Ministère de la Culture / APOR 156776)

concert of complaints and groans. We rose, we fell back, we rolled... Among the evacuees were a few, grievously ill, who were lying on stretchers. They did not have the strength to complain, but one could read such suffering in their features that it moved one to pity. I myself was chilled through, wet to the bone. I tried to stand up to go to them, but—bang!—a sea would strike and I found myself once more on the deck.

We kept leaping, and our boat seemed lighter and lighter. The captain, a very brave, round little man, shook his head... The sea would remain rough for the whole blessed day...

On the horizon, there was no hope of a change in the weather. The captain was right; one had to resign oneself to it. But just try resigning yourself when a frenzied dance shakes you from bottom to top, top to bottom, and so on.

The sea, all white-capped, had great troughs into which we disappeared, only to climb up again on the other side. Then, over again, over and over. A destroyer in the distance sped by, riding the waves at a magnificent pace. As for us, we were making six knots. The silhouettes of cargo boats, pitching majestically, were outlined against the sky...

In the end, it was decided to huddle us together in a great heap. We would be warmer that way. The captain generously covered us with a tarpaulin. I will not say it was very pleasant underneath, but we were so weary that we no longer thought of moving. We pitched, we rolled; there were cries, groans, and the sound of retching. The sea washed away the excess, the wind swept away the noxious emanations. We lay one upon another, all of us wet to the bone...

The hours passed in this dreary distress, and so we came to the head of the Mudros roadstead. Sheltered behind the island, the wind no longer blew, unless it had fallen as if by enchantment. The sea had suddenly become calm again, and a little mist spread along the horizon.

As if by magic, everyone had woken. Even the sickest men were stirring. We were no longer under the tarpaulin, but standing. Cigarettes were being lit. It was like the waking of a brood of chicks. Life returned; we felt revived, breathing easily, thinking of nothing now but gazing about us. Truly, changing from one minute to the next, it was a curious spectacle. I myself had completely forgotten that my white dress was clinging to my back. I felt full of energy, despite the great fatigue that made my legs a little weak.

But our entry into the roadstead was worth watching. At nightfall, the ships put out, all lights doused, standing out to sea to follow their appointed courses. That evening, there were fully twenty of them, all large vessels, the most massive of which was the *Olympic*.[1] Her formidable bulk

1 British ocean liner, sister-ship of the *Titanic*, requisitioned as a troopship.

moved with ease among the others, while we—lost in the throng—had to manoeuvre with skill. A blast or two of the siren, depending on whether we passed to port or starboard. Then, with a skilful flick of the helm, there we were, almost under the nose of one of these behemoths. We dodged the danger, then pressed on with renewed vigour until we had passed them all. Never shall I forget the impression it made. We, so small, lost amidst that horde, darting alongside them, pitching and sliding—we, for whom they seemed to spare not even a glance...

All those ships standing out to sea appeared to me that evening like a flight of nocturnal birds, waiting for the night to take wing and fly far away in search of their prey.

———

PART IV

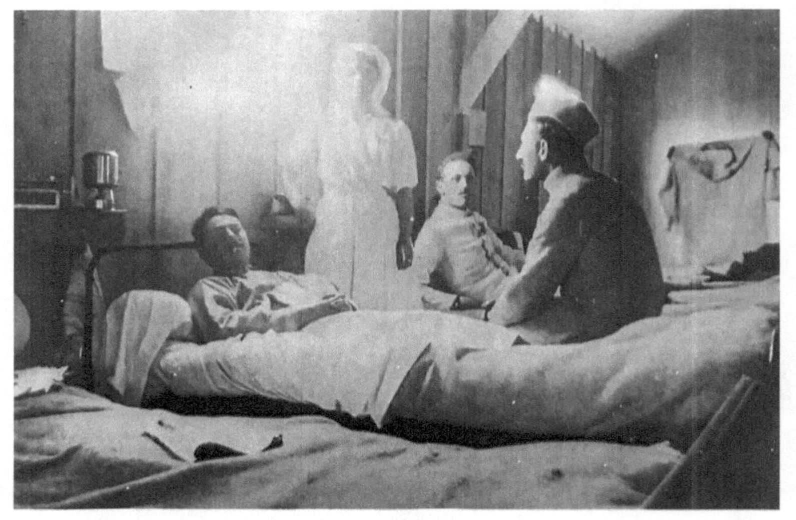

Interior of a ward at Evacuation Hospital no. 1.
(Elisabeth Jardin Collection)

I

Mudros, November 1915.

A morning heavy with dense fog. A raw, icy winter morning. The banked mists—that great white shroud familiar to northern lands—weigh down and penetrate. One can discern nothing of the things closest at hand, and silhouettes are lost, cut down as if by a shutter. All is seized, diminished, effaced.

Shadows loom up, bringing with them a sort of irrational fear, an almost morbid dread of this severance from the visible world. All around reigns the great sadness of instinct gone astray, lost. We speak in slow, low voices, in broken phrases, with the intolerable sense of the unknown, of the impossibility of fighting it.

Our camp, this day, breathes uneasily. The tents seem to spring abruptly from the ground, like immense snow-flowers. Other tents, all pale, sag a little, somewhat withered.

And all around, within the immense silence, is the immense noise coming from the open sea—from that 'out there' where distances can no longer be gauged, and of which one knows nothing. The sirens moan, weep, and implore. There are long sobs, endless wails, hoarse like those of despair;

cries of rage and of summons, brief and deliberate, plaintive as a child's; peremptory cries that lash out, and the long, agonising wail that bursts forth like a tearing wound. The cries of all the ships together, the infinite distress of the lost pack.

To-day is a day of war, and every ship knows her responsibility. They must—despite the elements—they must press on, obey orders. The struggle is more tragic, more bitter. The sinister concert continues its chords. The atmosphere is saturated with it, retaining the shrill notes like tendrils that pierce the ear and squeeze the heart. And once more, the long, agonised cry, stubborn and unyielding, bursts forth like a tearing... and comes again...

Oh! These moans, these wails, these supplications!

———

II

Mudros, November 1915.

There is misfortune in our hut. One of our nurses has had to be invalided back to France on account of a severe intestinal fever. And now our Matron, *Mademoiselle* Oberkampf[1]—who subsequently refused to be evacuated, remaining faithful to her post—is confined to bed in her turn. There was talk of typhoid. They were not mistaken...

Two of my orderlies have been struck down with it as well. The ill wind continues. It is a squall that will perhaps pass, but there are so many of them in the meantime...

————

1 Ms Yolande Oberkampf (1874–1954), a nurse with extensive experience (notably in Morocco 1907–13). Requested for Mudros by a medical officer familiar with her competence, she served as Head Nurse. In October 1915, despite contracting paratyphoid, she refused convalescence to continue working. Served later at No. 7 Auxiliary Hospital, Salonika. Awards: *Chevalier* of the Legion of Honour, *Croix de Guerre*, Royal Red Cross (UK) and several mentions in despatches.

Morane-Saulnier monoplane of *escadrille* MF 98 T,
the French army air squadron based at Tenedos.

(François Masnou, Agence Rol)

III

Mudros, November 1915.

Imagine a morning of war. Imagine this land of the East where nothing in the soil recalls France. Imagine a grey, damp morning. Imagine the camp all slumbering in sadness. Then imagine the great golden ray that lashes the clouds, scattering them for its own greater glory...

Imagine all this, and then that awakening of every man and every thing... Then look... you will know... you will understand more.

In the air, the uninterrupted hum, a hum that grows distinct and pervades the atmosphere... A strange noise that swells... and swells... The double sound clashing against itself—the sound of aero-engines shaking the empire of the skies.

You start, recognising that sound which calls man from his lair, that sound which puts the beasts to flight... which compels one to look up, high, very high, ever higher...

You thrill with emotion... You forget many things, and you long to see... You fall silent to follow with your eyes those magnificent birds, those created by the hands of men.

They came from the North... Their outlines could be guessed at; they approached with the fiery rage of a mastering will... They braved the turbulence, they seized the immensity of the skies... The magnificent birds, in the clarity of an autumn sky, wheeled above the camps, wings outspread as if for supreme victory...

They pursued their circles... they vied in ardour... They grazed one another... They leapt in that deliberate encounter. Perhaps they whispered the cabalistic words that intoxicate human strength?... And turning without respite, gliding in long, supple movements, skimming the ground then climbing again in a wide sweep, they departed to where men's eyes could scarcely discern them.

They drew behind them, fastened to their tricolour cockades, thousands upon thousands of eyes... Chests heaved, throats tightened, and voices fell silent. On earth, nothing was heard but the mute trembling of emotions intensified tenfold...

It was the greatest awakening in the camp...

Each soldier felt within himself a new strength, nobler aspirations. In a moment of divine understanding, he dreamed of the shared sacrifice made for the nation, of which he was the custodian... He forgot his own part to dream of other renunciations. He experienced individually that mass rapture which leads man to his apogee... He had a still clearer vision of creative ideas, ideas that would regenerate the world and lead it to triumph...

It was the greatest awakening in the camp...

The birds sported in the air, testing their mysterious strength. Steel breastplates glittered, and the armour made necessary by the greatest of wars turned these human birds into terrible birds of prey... The fuselages, like ribs and muscles, cast swift flashes... And the men still questioned those who came from further afield than they... They questioned them with their soldiers' eyes...

―――――――

Filing past the nurses, what appears to be a column of Turkish prisoners of war.

(Elisabeth Jardin Collection)

Nurse Jardin with two local children.
(Elisabeth Jardin Collection)

IV

Mudros, November 1915.

Oh, that retreat from Serbia—how we lived it...
We knew many of those who had gone out there.
We knew of the troop movements, and we were not
unaware that they were fighting one against ten. Yet
I must say that not for an instant did confidence
waver... We kept faith to the very end. Nevertheless,
there were hard moments to endure... Suddenly,
on our island, it seemed we were abandoned by
everyone. The men were leaving; reserve supplies
were being shipped to Salonika. It was a frantic
business that overwhelmed the French pier... And
then the ships in the roadstead grew fewer... Life
became calmer, more uniform; the mails became
rarer, and we had the impression of being cast-offs,
forgotten...

———

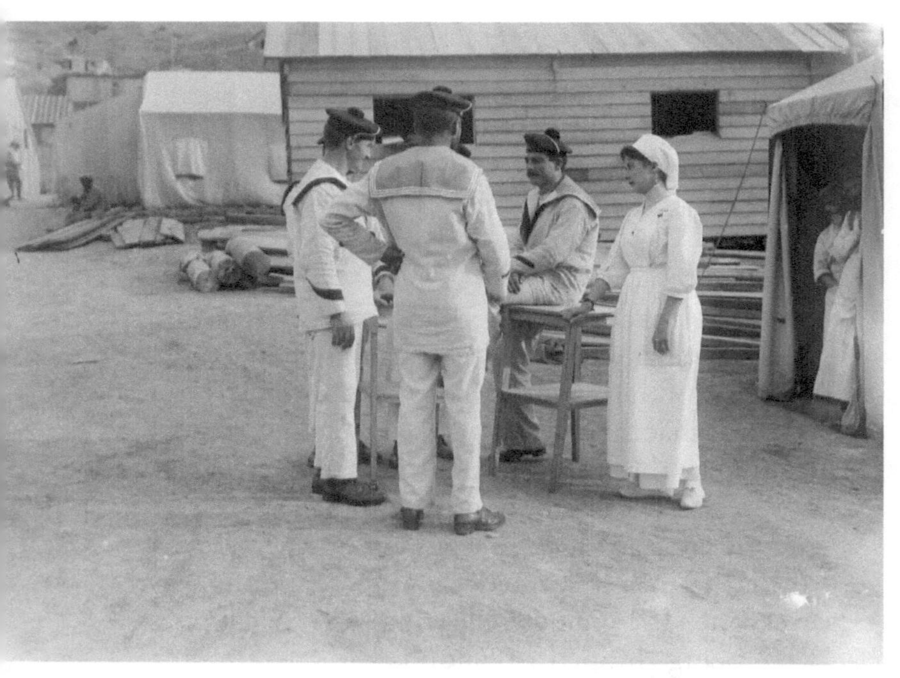

French sailors with a nurse of Evacuation Hospital no. 1.
(Ernest Brooks, Admiralty official photographer)

V

Mudros, November 1915.

Bah! One must record, one must tell what one knows, what one has seen that is lovely, very lovely—like what follows here. I wager that no story will ever move you more. This story is not very complicated. Indeed, it is scarcely a story at all; a fresh and beautiful story, a very simple story.

In our own camp—so vast and filled with wooden huts, *marabouts*, A-frames and others; in our camp, where some meagre oats struggle to grow—there are crowds of soldiers, great gaunt devils of soldiers, sick and emaciated.

Above all, there are the long fevers that wreck a man, often cutting him down beyond recovery. Those fevers are terrible; they strike when one least expects them, and I forget what scientific name they have been given...

These sick men, and other patients too—very serious ones, you understand—have each day, listen to me well, their ray of sunshine...

He is a sailor, a boy, a little fellow of nineteen who wears the stripes of a Leading Seaman. A beardless face, a little rough, with large black eyes, guileless and distant—eyes that feel and eyes that love...

So, on the days when they are not out scouring the seas or bombarding some coast, on those days Conort (that is his name) brings his comrades along with him—sailors, naturally. They arrive gaily, hurrying down the great plain, where their swaying, all-white silhouettes gleam like lights.

And then, as if it were the most natural thing in the world, they bring with them provisions of all sorts: one with tobacco, another paper, oranges, lemons, sweets, books, newspapers, and so many other things procured at great cost in money and imagination...

And each one bends over... each one distributes...

But it is Conort who leads the way; it is he who knows the sickest, the poorest, the most forsaken... It is he who writes to the families in the evening when, back on board, he has finished his duties... It is he who divines, who knows so very much, that all his comrades follow him and imitate his example...

Conort maintains a vast correspondence, a voluminous post... He has made it his task to write for those who cannot... He has made it his task to reassure the mothers, the wives, the sisters...

There is no need to know Conort well to feel at ease! His manner—so open and frank—of approaching a sick man, his gentleness and his delicacy, indeed his very grace, make him the friend... the confidant... And Conort has his sailor's smile too, full of quiet pride, sweet serenity, ardent bravery, and magnificent devotion...

And I swear that you would envy those who glimpse him when he is unaware... He finds the right words, for he speaks from the heart, and his face bends down, full of solicitude and tenderness, over the one in pain. Contagion holds no fear for him... He wishes to see those who are isolated... Indeed, he prefers them... He has a way all his own of restoring failing courage... He consoles, he reassures... and he cheers...

Conort has become a very dear friend to all, and it happened at once. And through habit, too. For consider: for months and months now, Conort has made it his mission to succour his comrades... Not a single holiday, not a single rest day... Whenever they are not under sailing orders, Conort arrives, cheerful and laden with a thousand good things...

For Conort is the delegate of the *Suffren*, of the entire crew of the *Suffren*—of that crew which every month gives its mite for the hospital... From the lowest to the highest, each vies with the other in generosity... Each man wishes to give more than he can, and each dreams of what more he might give. A sailor's generosity, unsuspected, made of such delicacy and kindness...

Conort's eyes are too misted with gentleness not to understand what others of his age would not. And Conort loves... loves... He loves all these soldiers who are, in a way, his brothers; he loves his ship, of which he is proud; he loves his officers with passionate admiration... and I swear that for his

Lieutenant, Conort would give his life a thousand times over, if he had a thousand lives...

There... My story is not quite a story; it is scarcely a portrait... I should have liked to tell it better. I should have liked to say more of the moral and physical succour that these sailor souls—these sailor hearts—bring to our soldiers so far removed from everything... Conort is in great part the *Suffren*, and the *Suffren*, for all who have passed through here, will remain the *Suffren*: a ship where they loved the poor devils in pain...

The other day, returning from the hospital, Conort said to his officer: "Sir... This is one of the finest moments of my life..."

———

VI

Mudros, November 1915.

This is another tale, but a true tale, so sad.

It was on one of those afternoons, at the hour when night approaches and all noise fades as if by instinct to make way for contemplation and silence, the hours of the vigil become longer, more peaceful. Each listens to the anguish or the hope within! One lives closer to the earth. One lives in more direct communion with all human suffering. One realises better the vanity of passing things. Sensibilities sharpen. One hears more distinctly the thousand murmurs that silence bears along... It is the great calm in which all brutality dies. And one has but a single desire: to preserve this marvellous inner peace. One loves this divine hour with all the keenness of memory reasserting its rights...

To-day, the softness of the evening makes the silence hovering over the camp more penetrating, and the three men sitting motionless beneath this tent, lit by a poor lantern—these men do not speak. They live strangely isolated from one another. Each, no doubt, stirs within himself those emotions belonging to the past that return so strongly with the tranquil beauty of Eastern evenings.

I take care not to disturb their silence. The

minutes slip by smoothly. Only, sometimes, one starts, because it seems that, in the night, footsteps are disturbing the stones...

I have been gazing, so calm, so gently calm, at this setting that surrounds us. The hanging ropes which, above the dusty, chipped tent-pole, dangle unevenly. The wooden table on which I lean, a table roughly hewn, made in a few hours, and all rickety. I notice that, by too close a contact, it is wearing through the canvas of our marquee...

And I muse on our remoteness... It is true that, back there in France, one could scarcely picture this group. These men keeping watch, so far from their country, yet so close to all they hold dear... I feel a sense of trust...

Weeks, then months have passed... One lives amongst them, quite free, breathing in a sense of holy protection all around...

The minutes slip away in the great silence of the camp as it falls asleep... But a man appears at the entrance to our tent. Then, without asking permission, he takes a half-open crate, turns it over, and sits down.

He speaks in a low voice, as though not to interrupt our reverie. He happened to be passing not far away when our light attracted him. He has come to see how we are getting on. We had had no communiqués for several days; did we know anything? ... A merchant ship sunk by an enemy submarine—a ship of little importance, mind you, but all the same there were deaths. And that agony, in the open sea, in the pitch-black night, took on a greater magnitude for us...

Now, our man remembered... And in a voice lower still, while the silence grew all around, he began his tale:

"One evening—or rather one morning, for it was two o'clock—on the 12th of May... I was at Sedd-el-Bahr, busy recovering a wounded man... The night was superb and there was silence everywhere when the guns stopped sounding. Even my poor little lad was not complaining..."

Here, he paused...

"The night, as I was telling you, was superb; it was full of stars, great stars drawing closer to the smaller ones as if to protect them. Each of us went about his task without losing anything of what surrounded him...

"All at once, a tremendous noise, an immense sound, a terrible explosion that crushed the very air... At first, we did not understand... Then softly, like the wail of a child, a cry rose from the sea... Other cries followed, melting into a single cry... *Ooh... Oo-ooh... Oo-ooh...* A horrible clamour, a desperate clamour, like the cry of all the dead united. It was terrible and mad... The clamour increased, filling the whole sky—the clamour of men fighting for life... Oh! It was intolerable...

"In the darkness, we could distinguish nothing, we saw nothing. Only the great cry kept rising, more

tragic, more frightful... more powerful than ever...
Oh, how that cry rose... Oh, how it endured...

"Then, as if a valve were being pressed, the
atrocious clamour sank, diminished slowly, more
painful and more terrible. It went on diminishing...
And that *ooh... ooh...* became nothing more than a
death rattle... One by one the cries died out, then
silence returned... Once more a long, long wail, an
immense appeal, a choking sob... and it was over...

"Even so, the night retained something like a sob,
and for a long time we heard within us that sound
of death... Our ears and our hearts were filled with
the sinister clamour... Oh! That night, no one slept...

"At daybreak, I went aboard a British destroyer. I
told them of our anguish... The officer, a friend to
whom I spoke, replied with a salute: 'It is the *Goliath*[1]
that has gone down; six hundred men perished...
But it is war, and so it goes—it is nothing...' And he
saluted again, as though to salute the dead..."

Our man resumed:

"All the same, it was rather grim..."

And he fell silent...

1 HMS *Goliath*, a British pre-dreadnought battleship, was
   sunk on 13 May 1915 after being struck by torpedoes fired by
   the Ottoman destroyer *Muâvenet-i Millîye* while stationed
   in Morto Bay. The attack caused the ship to capsize rapidly;
   of a complement of approximately 750 officers and men,
   more than 500 were killed.

In our own silence, we heard distinctly, though from very far away, the pure sound of a thin bell... Then the *Parce Domine* followed...

It was the evening prayer reaching us on the wind, bringing us its sweetness... And everyone prayed with their heart, for the dead, for those who had died that night... And our eyes, filled with tears, instinctively sought, through the opening of our tent, a patch of sky where a few stars might shine...

———

(Elisabeth Jardin Collection)

VII

Mudros, November 1915.

Last night was hard... No one slept. The north wind blew with a harshness unknown to us. Never had we heard it so close. The tents shook like beasts, bracing themselves in an unimaginable rage. The canvas snapped, as if bearing a grudge against those it sheltered... The malicious, biting wind rushed in. The supports holding the tents shook noisily from right to left with brutal jolts. The guy ropes flew and clashed together. Everything trembled, exasperated in a frenzied and despairing struggle. And the wind kept mounting the assault in a perpetual hand-to-hand combat. Never had anyone seen such a night... And never had anyone felt so alone...

To-night will be just like the last. The same storm is raging this winter afternoon, with perhaps a little more violence. The same sky overburdened with heavy clouds, those same grey masses, with too many shadows... The same rifts, as if to measure their depth...

Only to-day, away to the north, great flashes of lightning sabre the horizon, cracking the darkness, spurting in great streaks...

This evening, in the camp, the tents are struggling as they did yesterday... Some seem unable to bear any more, ready to give way. The lamp lit inside gives them a ghostly look... To-night there is room for fantastic tales, for witches of the 'moors' and ghosts from far-off lands. And then, the rain has cast down its lugubrious misery... Everyone shivers, diminished by the chills... Everyone is cold, without protest... Only, this evening, a dream passes too: it would be good to be by the fireside, pressed close against one another, listening to some grandmother's tales...

Oh, how the wind whips in long, treacherous gusts; how it growls in the distance, and there— right up close—against the ill-fitting wooden door where the rain drives in... How it obstinately batters the roofs, between the corrugated iron sheets that let the water seep through... How it sings of death with this concert of moans, of gnashing, with all this wild frenzy...

The night is black, oppressive...

Oh, the wind this time screeches with greater fury... The gusts lash... the timbers crack... Everything cries out in pain.

That gust was the strongest yet! A door has been torn from its hinges... It rains in heavy drops and the wind rushes in like a whirlwind.

The men struggle in their turn. They struggle, frozen, feverish... The door is put back. It is shored up with huge stones. Nails are driven home, and the sound of hammers is lost in the storm... And the wind howls... howls...

In the wooden hut, on rope beds, beneath spread-out blankets and greatcoats, figures lie motionless... Black skins, for the most part, lost in the shadows... Two lanterns, meagre and miserable, swing in the void, casting their pale light about. No warmth comes from them... In the very middle of the room, two tricolours hang still...

No one speaks...

No one sleeps...

Everyone listens...

The wind continues to drone, passing in great waves, the better to seep into everything...

Two shadows have risen up... Two black shadows, one wearing a red cap, the other a white. Two shadows facing the East. They no longer see anything of their surroundings, pay no heed to the wind that howls... howls...

In a single supple movement, they have knelt, with the posture of idols...

Their foreheads now touch the earth... They rise again, and one hears long litanies recited in rhythm, words they enunciate religiously, with a reasoned, unshakable faith, detached from all human regard.

Oh, the wind... This time again, it gained the upper hand. The whole hut cracked, furious...

The litanies continue, slowly chanted in a harmonious tongue of strange cadences... The Marabouts[1] pray to their God: 'There is but one God who is God, and Mahomet is his Prophet...'

Men shiver with fever... It is cold... The night is black. The wind continues its macabre madness, its wild, unbridled dance...

Outside, silhouettes struggle... the clouds flee frantically; sirens moan, tireless, obstinate, and imploring...

One must find silence within oneself, and listen to the sad memories that crowd the smallest corners of the heart to-night!...

In the next room, a man cried out, called... The wind makes the candle that lights him flicker... The tiny, bewildered flame no longer knows which way to turn. It throws itself from right to left, straightens up, then flares even harder... Stays grate, and one hears again the wind approaching in great strides, forcing every obstacle...

————

1 Muslim chaplains serving with the West African units in the French army.

Looking north-east from the hospital, towards the cemetery.

(Elisabeth Jardin Collection)

PART V

Evacuation Hospital no. 1, Mudros.
A good view of the terracing, with the town in the background.
Jeanne Antelme is likely the nurse at the top of the stairs.

(Elisabeth Jardin Collection)

I

Mudros, December 1915.

On a fine winter's evening, on a very mild night, one goes out there, following the path that slips between two hills, bordered by a ravine carrying slow waters, full of dreams. On this path, every rustle, every jar, every footstep lingers, slowly echoes. And there are so many shadows that one never sees the end of them. Thus, many hesitate to follow it, and that is why, I believe, one has it all to oneself—this path that one loves, that one loves for its silence and its solitude!

With the coming of night, the stones that have torn themselves from the hills and that jostle all around—bristling, straining, arching, rising, and stretching out—those stones clothe themselves in the fantastic. The night magnifies them immeasurably, and the heaped-up shadows adorn them with mystery, with allure. And then, when a fair moonbeam descends upon them, one sees them come to life, one by one. They listen to the subterranean water singing at their feet, for them alone, the eternal and troubling chant, the sweet melody of tranquil evenings.

Instinctively one stops, and in the great silence of the nights one hears, one by one, the frail or deep notes that rise from beneath the stones and whisper in one's ear the marvellous songs of the heart.

Listen to those tiny cascades! That small, immense sound—deep and melodious—filling the air!... Listen to that gurgling, that perpetual resonance of waters meeting! Listen to the other sound, the sound of the tiniest echoes swelling!...

Sometimes the light hooves of a donkey disturb the stones on the path. He comes along all muffled in wild heather, and one can scarcely make out his gentle, drowsy head. "Kalispera!" says his motionless driver, in a singing and melodious voice, dwelling at length on the penultimate syllable. "Kalispera" is repeated, heard again as other little donkeys file past, and "Kalispera" takes on in the night the charm of a great passing dream.

And then, at other times, it is the crows—huddled together, gathered—crying out some litany or other; and their throats, restrained and muffled in the night, speak an unknown language. They tell, no doubt recounting to one another, the winged legends of their desires of a day. They tell, they tell again, they tell for ever.

Then, it is night once more. Silence, and the water going *glug-glug*, which, dropping an octave, accompanies the light song with a deeper note— *glug-glug... glug... glug...* And great stones hide the spring!

And then, higher up, the flock scattering along the hillside; the slow bells alternating with the silence, fading, returning, falling silent again only to resume, now distant, now near. Pensive bells— *drong... drong... drong...* Hesitant bells— *dring...* stopping, then *dring...* making up their minds. Bells that jolt and clash and ring all at once: *drong... clunk... dring...* Delicate bells, loving and solitary, full of a very sweet song; *ting... ting... ting...* they say, dwelling on the final note. The great flock of white sheep with long silky fleeces drifts in its own dream, at random, in the night, into the fading night.

An icy laugh or a heart-rending wail pierces the darkness: the owl laughs or weeps at being a bird of the night...

And once again, the great peace of evening descends. And once again, the slow, loving, solitary bell repeats its song—*ting... ting... ting...* It is the dream returning. It is the dream of an evening. It is the dream of a night.

———

The French section of the military cemetery at Mudros.
(Elisabeth Jardin Collection)

II

Mudros, December 1915.

To-day, the funeral of the chief physician of the convalescent hospital—the one right next to ours—took place. Doctor A— had come to our hospital, where he died of typhoid.[1] Yet another who had done his duty brilliantly over there, on the peninsula! He had been in the breach from the beginning of the campaign. And then, owing to sudden exhaustion, he had taken up duty at Mudros two months ago. He was a tall, strong man, solid as a rock... And once again, that damned fever had the final say. He was very ill, then better. We thought him saved—a relapse finished him...

1 *Médecin aide major de 1ère classe* (Lieutenant) Jérome Ferdinand François Arnoux, died of sickness, 16 December 1915, age 41.

'To the *Suffren*. Mrs Laporte and me.'
(*Elisabeth Jardin Collection*)

III

Mudros, December 1915.

The other afternoon, a south-west wind was blowing, thrusting the colours flying from the mizzen gaffs of the ships in the roadstead right before our eyes. Ours—my God, how they streamed out! There was no mistaking them. The blue, the white, and the red—they positively stung the eyes. We were not a little proud, and not a little moved, either. For our colours were the finest, to be sure. It is true that they knew how to command respect... Ah! that blue, that white, and that red, how they gripped the heart... And I assure you, one must see them in so pure a setting to know what they can suggest to you. No... you see, one has the soul of a child and the vigour of a trooper before that banner...

And then again, one thinks in spite of oneself of all the blood shed, of all the sacrifices... And one tells oneself that our colours owe their proud billowing to those very sacrifices. Thus, in revering one's colours, one reveres one's dead; one pays them their just homage.

Yes, in the great roadstead, in that vast amphitheatre of mountains with their harmonious, supple lines, in all that mauve of the twilight, our colours stood out in sharp relief... And one could look higher still...

In a long gust of wind, we heard the sharp crack of blank revolver fire, then the vibrant voice of the bugles sounding as they do every evening... Snatches of the *Marseillaise*, drifting across the water, reached us; and slowly, reverently, we watched our colours leaving the mizzen gaff...

They were lowering the colours... The bunting glided down... No one moved... We listened and we watched... Then everyone went their way with the three colours still in their eyes...

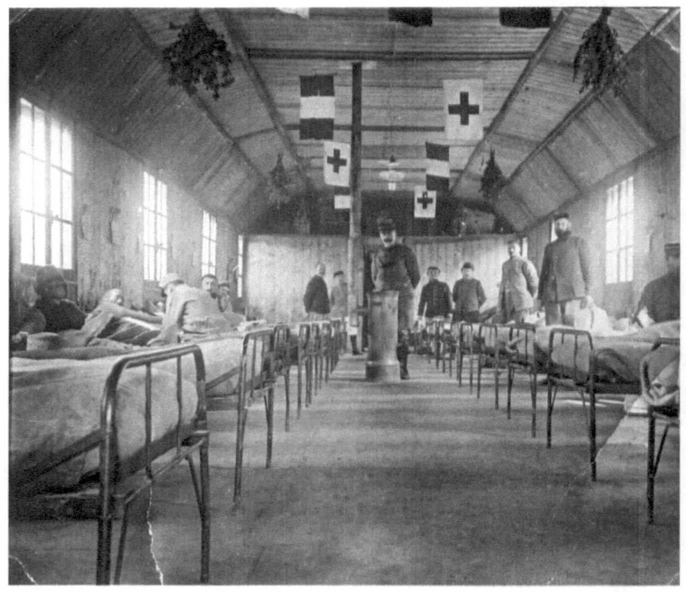

Ward interior, Evacuation Hospital no. 1.

(Elisabeth Jardin Collection)

IV

Mudros, December 1915.

Yesterday I received a visit from Admiral C—,[1] the British Admiral commanding the naval base at Mudros and Governor of the island! He is an old friend from my childhood... I was also told, with a wealth of detail, of the raid by one of the many British submarines that penetrated the Sea of Marmara. I noted the details down to remember them better, and also to retell them to my patients...

It concerns—listen closely—the submarine *E11*, which departed on 19 May for the Straits. It is a difficult story to tell, but I shall try all the same... Let us begin. Scarcely in sight of the Straits—it was about 2.45 a.m.—she made towards them. The weather was fine, the sky very blue... but that mattered little, did it?... for they had to dive immediately when abeam of Achi Baba. They dived to 25 metres to slip under the mines; three hours later they were rounding Kilid Bahr, having touched bottom which was only at 16 metres. They continued calmly on

1 Rear Admiral Arthur Christian, Royal Navy, second-in-command of the Eastern Mediterranean Squadron. (The base commander and governor for most of the Gallipoli Campaign was Rear-Admiral Rosslyn Erskine Wemyss, who departed Lemnos for England on 23 December 1915.)

Rear Admiral Arthur Christian *(left)* meets with the nursing sisters of Evacuation Hospital no. 1. Standing beside him is the matron, *Mlle* Oberkampf. The author, Jeanne Antelme, is likely the nurse in the centre of the photograph.

Elisabeth Jardin *(left)* meets the Admiral.

(Elisabeth Jardin Collection)

their course—'calmly' is perhaps excessive, for the concentration required defies all imagination—when the periscope was spotted by a vessel and enemy destroyers... So, doubling their speed, they set off at full tilt towards the north... They did not stop; they continued their course, rounded Nagara at a depth of twenty-five metres and passed Gallipoli seven hours after their departure. A little more than seven hours had sufficed to bring them there... but they proceeded still submerged, they did not come up; they carried on and three hours later, to rest, they lay on the bottom near the north coast...

Finally, at 9 p.m.—that is to say after nineteen hours submerged—they rose to the surface... Look out... the hunt was about to begin... but scarcely were they out when they had to dive again twice in succession, enemy destroyers having sighted them... That did not prevent them from making the regulation signals to their Admiral to tell him of the success of the enterprise... Wasted effort... The Admiral did not answer, and for good reason: an antenna wire was broken...

So, every day, submerged or on the surface, they continued their mission. The Sea of Marmara belonged to them. They came and went. They surfaced, they dived... On 23 May, they had the good fortune to spot a German gunboat off Constantinople... Great joy: they moved into position and from the port tube they torpedoed her and sank her... The torpedo struck her on the starboard side just amidships... But the gunboat did not consider herself beaten; while she was sinking, she spotted the periscope, at which she opened fire. The first salvo hit it... They had to make for the north, towards the island... After having surfaced, they repaired the periscope to fit a new head...

But they were not always underwater; when the weather and the enemy permitted, one could think a little of oneself too. So the crew bathed sometimes, though not as often as they would have liked.

On 24 May—yes, that is right—they were making course to the north-east, when they spotted a small steamer heading west. They first identified her with the periscope, then, surfacing abeam, the commander of our submarine gave the order to the steamer to stop... No result... She continued to steam; so, to force her to stop, they had to fire a volley of rifle shots at the bridge. And the crew was ordered to abandon ship... Oh! The crew did so with terrible haste... Almost all the boats capsized in the panic. But happily there were some left, which allowed the men in the water to be picked up...

But the most curious thing was the appearance on deck of an American gentleman... Very correctly, he informed the commander that he was Mr S. S. of the C. S.[1] and that he was delighted to make his acquaintance.

He was loquacious, and so charmed to see them that he told them this steamer was transporting Turkish

1 Mr Raymond Gram Swing, from the *Chicago Daily News*.

marines to Chanak and that he was almost certain that stores were to be found on the boat... Immediately they went alongside the steamer, and one of our officers went aboard with a demolition party.

They found at once a 150-mm gun lashed to the fore hatch, and in the fore hold a large 150-mm gun carriage and several mountings for small 12-pounder guns...

As for the after hold, it was full of 150-mm projectiles upon which rested about 50 large ammunition cases marked Krupp...

A demolition charge was immediately placed in the after hold in the middle of the ammunition cases... Need I tell you that this made an enormous explosion? A column of flame and smoke, then the ship sank... At that moment, they spotted smoke in the east... Immediately, they dived to attack. But the ship, having spotted their manoeuvre, changed course... So, in their turn, they surfaced and pursued it... It was a supply ship... They fired their torpedoes from port, the ship was struck just amidships; immediately fire broke out...

The next day they spotted a transport which they sank as well.

Two days later, they dived unseen off Constantinople. Yes... off Constantinople. Another large transport was berthed in the arsenal. Immediately they fired a torpedo from the port bow... This one failed to run... They fired a second from the starboard bow... They saw at last the track heading straight for the large vessel… But at that moment, they could not tell whether they had hit their mark, for the submarine was swept towards the shore by the current…

To stop rising, they had to go full speed astern while flooding the internal tanks. Their heading shifted from south-south-east to west, passing through east and north. The commander concluded that they were grounded on a bank under Leander's Tower, and that the current was swinging them round… Having then headed south, they went ahead and the ship passed gently along the bottom, sinking to 25 metres… They touched again several times at this depth, then they lifted off the bottom, and, after returning to the surface twenty minutes later, they perceived that the entrance was cleared… On 26 May, they treated themselves to a rest day in the middle of the Sea of Marmara…

And every day, they continued their hunt, sinking transports and supply ships… But life on board the submarine was in reality trying; the air became unbreathable due to the great quantity of dirty linen, and fresh water became so scarce that they had to limit the frequency of their washing… So they sometimes spent the day right in the middle of the Sea of Marmara "to give the ship and men a good clean"…

The weather continued to be magnificent at the end of May. The moon was resplendent, which prevented them more than once from pursuing their hunt.

They continued this existence, on perpetual alert, until 7 June, when they made their exit... They had to dive to 27 metres off Gallipoli, but immediately after rounding Kilid Bahr, the boat's trim was no longer the same. It became erratic, which necessitated an increase of eight tons of water in order to descend to 28 metres.

Two hours later, a noise was heard as if they were touching bottom... Knowing that this was not possible, given the depth where they were, they rose to six metres to see what it meant... It was a great, big mine, which was about six metres from the periscope and doubtless snagged by its mooring cable on the port hydroplane... They were dragging it in tow...

They did not think of freeing themselves; it was impossible because of the shore batteries, so they continued to make way through the strait at a depth of nine metres.

An hour later, they rose to six metres, beyond Kum Kale... There, they decided they must abandon the mine. They went full speed astern after emptying the after ballast tank to let the bow sink and bring the stern to the surface. The speed astern and the current cleared the mine from their bow... All was saved...

It is very simple, is it not...

The nurse on the donkey is likely the author, Jeanne Antelme.

(Elisabeth Jardin Collection)

V

Mudros, 24th December 1915.

Ah! Master Jean, you did not know that one day you would pass into posterity, and what posterity! No, for sure, you had no inkling of it. And yet I had promised you, and more than twenty times at that. Oh! If you were to read this, how you would carry on; but I swear to you, Master Jean, that truth alone shall issue from my mouth.

Why hide, Master Jean, why flee? I assure you that your example would be salutary. And I do not see why, because it pleases me this evening to tease you a little, you should take umbrage.

We all know, Master Jean, that when it is a question of fulfilling your ministry, you would put any prelate to shame. But it is precisely for that reason that I wish to introduce you to all those who do not know you!

Master Jean is the priest of X—, situated in the Basses-Alpes. He is short, plump, round-faced, dark like any good Marseillais and cheerful as a true Southerner.

Ah! Bless him... when he plays *boules*, you should see the concentration he applies. He bustles, leans, takes his run-up and—wham—tries to hit the jack. Do not ask him the score... He will answer you with conviction:

159

"It is a cut-throat game..."

Ah! Yes, Master Jean, at the hour of rest, it is at the *boules* court that one finds you. There, with some dozens of your compatriots, you replay each day the match of the day before...

But that is not all, alas! Do you remember? Yes, do you remember the other afternoon? ... When that British airship manoeuvred above the roadstead,[1] remember that cry which burst from your heart. You had put on your finest accent that day. No, I shall quote only your august words... your exclamation: "I wonder all the same how they manage to live in that great box." And hands in pockets, nose in the air, you gazed not at the nacelle, but at the great steel cigar. Afterwards, you apologised... You explained to us that at R—, in that small commune where you assumed the heavy responsibility of a country priest, "one did not know of such animals."

And then again, remember... The other day, when one of those aeroplanes that inhabit Tenedos came to pay us a visit, remember, nose still in the air and hands in pockets, what you said:

"My dear... how they pedal..."

Yes, Master Jean, we know so many other stories of this kind. I shall not tell the one a colleague of

---

1 The Royal Naval Air Service operated a non-rigid airship during the Gallipoli Campaign. Its envelope was coated in aluminium dope to reflect sunlight, giving it a distinctive silver sheen. The gondola consisted of a repurposed two-seater aeroplane fuselage with the wings and tail removed.

yours told me, because, the other day, I saw you turn quite red...

I shall content myself with this one... Everyone knows your character... One has never seen you angry... so you have only yourself to blame. You remember... the other time—you always need reminding—Yes, when, in a teasing mood, I asked one of your *confrères*:

"Do you think Master Jean could have made a Jesuit?"

"Master Jean? You might as well ask a butcher to become an artist."

Oh, Master Jean, how very quickly you answered:

"Back home in our country, when a priest goes bad, we say he has turned Jesuit..."

I saw then that I had blundered, and I tried to recover myself. So I asked you to attend to a sick man, Master Jean, and you said not another word... You simply left.

Anyhow, Master Jean, we must admit it: you are slightly—oh! very slightly—but you are a gourmand. It is true that at Mudros one might easily become one. But... but you speak with such a tender voice of pâtés, of wines, of rabbits and hares; and even—which is more serious—of the truffled turkeys your parishioners offer you. You even promised to invite me one day to these feasts.

It is also true you added that, to earn all this... you remembered your calling as a priest; and that every morning, Mass ended, you went across the fields to find the poorest, those who could not afford a farm labourer... You took the labourer's place. That was good...

But all the same, you are a gourmand... Do not deny it, for besides the poetic nickname you were given—Father Joy—another has been added (and by your patients, no less!): that of 'Father Toddy'! Oh yes, you sample your patients' toddy. I do not say you take their share—God forbid!—but you come to an arrangement with the dispensary orderly so that he is not too stingy... It is a sin, a grave... a great sin!...

Your patients adore you, I grant you; they are ready to be cut to pieces for you, and the sight of you makes them smile. But... but... you are a gourmand.

Take, for instance, this menu you amused yourself by composing... The words caressed your palate, which is why you chose them so... otherwise... for as a meal, it was rather meagre. Listen to the announcement of the dishes. Look above all at the letters composing the word MENU. They are too large; that alone suffices to condemn you. So, let us see... MENU... and then Entrée... sardines... dry-cured sausage... butter... It is too much. Let us continue: Entremets, pâté (La Savoureuse brand); Main Course, smoked ham, Mudros asparagus... *Mudros asparagus*... And notice the prominence you have given your entremets... I continue: DESSERTS, in immense letters... almonds, fancy biscuits, Alpine honey, Var nougat, fruit cake, Ardèche chestnut cream. Wines... Bordeaux, Graves, rum... cigars...

Now that is a wartime dinner.

Master Jean, I told you the other day that if I could send you to confess to Father X—, it would give me great joy... You replied:

"Oh! It is I who hear his confession... and I give him a good talking to, I assure you..."

At once I felt a great respect for you, and put on my sweetest voice to suggest:

"I hope you are severe..."

You answered:

"I try... I try... but what can you do? He is sincere... I do tell him he is a nuisance sometimes..."

Ah, Master Jean, that is an ugly word... Let us change the subject, if you please...

Well now... after all this talk, I shall tell you what I really think of you. Do not get angry... and do not blush either. Well... I say to you: thank you. Yes, thank you; for you have been tremendously devoted to your patients; you knew better than anyone how to care for them. You never spoke to them of religion, and, believers or unbelievers, you embraced them all in the same sympathy. Well, Master Jean, that is worth knowing. In life, you see, one must learn to be tolerant, one must respect the convictions of others, one must respect all liberties; and that is the best way to make your religion loved. Do not forget that a free-thinker can be the most honest man in the world, and that to have a conscience, a creed is not enough. God belongs to everyone.

Besides, you know all this, and we have spoken of it together...

Remember, Master Jean, the other evening? It was seven o'clock when I sent for you... It was Christmas Eve. I had seen all my patients. In my little office where the wind gets in, between the two 77-mm shells brought back to me from Gallipoli, half a candle burned in a shell-fuse... You came... And I said to you: "Master Jean... I wish to make my confession." Yes, that evening, I believe I would have confessed to the whole world; it is so good to speak the truth and hide nothing.

You answered me: "Then we must be serious." And while you lost yourself in prayer, I—seated on my table, for my chair had been taken—and you standing, told you what I believed I had to say... It did not take long... You spoke a few words to me that were as much those of a friend as of a priest... And then came the absolution... And afterwards we spoke of a patient who was causing us anxiety...

Master Jean, that evening I knew an immense peace within me, an immense well-being... And when I found myself outside again, in the open air, with all my stars that seemed to greet me joyously and that shone, gliding by—Master Jean, that evening once again, I thanked God...

———

VI

Mudros, 25 December 1915.

A night calm and pure. A night so studded with stars that it seemed they had kept some mystic rendezvous... A night when peace slipped down from the heavens, when the planets grew large in their solemn beauty, when all became immutable. A night luminous with clarity and quiet reverence, a night when everything spoke to the heart of immense peace and infinite forgiveness!...

To-night, the soul has withdrawn into itself, in close communion. It is peaceful and serene, it listens and vibrates with love. There is no rancour, no regret, only a great peace! It is moved and trembling, for to-night is the solemn night, the night when God draws near to His creatures... The night when one descends to the very depths of human vanities, where all vanity disappears! The night when one loves the God of Love with a sanctified love! The night when one forgives, the better to love!

The night that bursts with love, that speaks of love, the night that calls to and celebrates the Creator, this God of love and truth whom we find in the immensity of the heavens! The God who says: "Love and forgive... Love!..."

Oh, this divine night! This divine atmosphere! Oh, this sweetness that steals into you! This great repose! This release, and this desire for love growing like a supreme will! Love and forgive! Love and understand! Love!... Oh! This peace of the soul that expands the soul and diminishes the body, until it bows it to the earth, the better to listen, the better to commune!

Oh God, to-night we kneel before Thee, the soul pure and freed from all bonds, and we shall hearken to Thy supreme laws of love!... We shall love with infinite love Thy sacrifice of love!...

Oh Christ, God of light...

It was under this Eastern sky, in this vast setting where the eye is lost, where everything sings in unison with our hearts, that *O Holy Night* and the simple carols of village Christmases rang out!... They are men's voices, nothing but men's voices, hundreds of men singing bare-headed, standing and respectful... The soldiers sing, solemn and reverent, and in their distant gaze passes the memory of other Christmases... We are no longer alone, since we are singing, and all those songs lead us back to the family hearth, to the soil that holds us by every fibre. Though we have before our eyes the silent fires of the neighbouring camps, the countless lights of the anchored ships, the long, supple silhouettes of the surrounding mountains, the soul runs, flies over the seas and finds again in some lost corner—deep in the Cévennes, or on the Breton moors, or further still, in some cottage of the North—a Christmas that is very much alive...

To-night is still a night of war; it is the sacrifice accepted, fulfilled; and it is the humble homage, the remembrance of all those who sleep for our peace! All those whose souls traverse the heavens and ask for respect for the cause for which they sacrificed themselves!

———

Elisabeth Jardin at Evacuation Hospital no. 1.

(Elisabeth Jardin Collection)

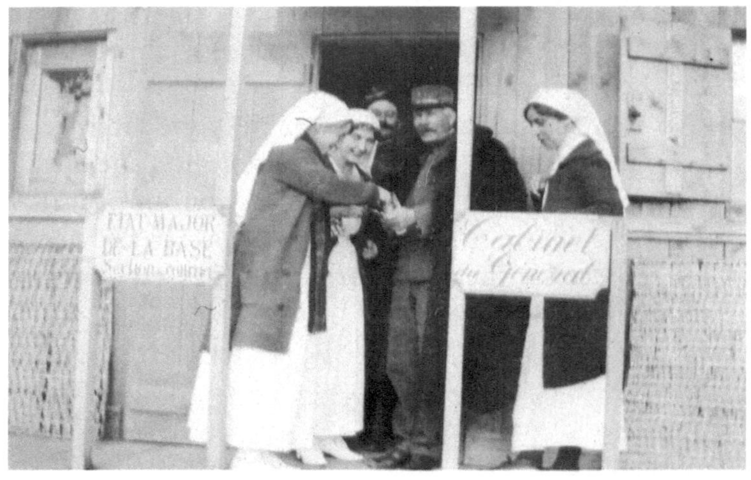

Nursing sisters with General Baumann, French
commander of the Mudros base.

(Elisabeth Jardin Collection)

VII

Mudros, 25 December 1915.

Our Christmas tree was a real success. We had scoured the whole island to find the makings of it... With all sorts of branches we managed to fashion it, and although it was of three different species, it still looked rather fine. We made bows of blue, white, and red paper which we pinned all over. It was thus quite dressed up, not counting the oranges and the prizes that adorned it. Standing high, it could be seen from afar.

Each patient received a packet containing a few keepsakes, and each had also a beautiful golden orange and a large mandarin. It was not much, but it was something at least—a thought, a way of remembering home a little... An obliging singer struck up the *Marseillaise*, after one of the soldiers had read us a most touching address. We all took up the national anthem in chorus, and Admiral Jaurès arrived at that very moment... This gave the singers more heart... He saluted at length, the Admiral, and his eyes were moist. He saluted the anthem and the men. Then he came to each of us and thanked us— for what?... Good Lord... That admiral was awfully popular, because we knew him to be very capable, and also because, off duty, he endeavoured to efface all distance between himself and his subordinates. He was no longer the Admiral, but the friend...

So our tree was a success, a real one... Then it was the turn of the seriously ill, who also each received a keepsake... And again the singer went around the wards to perform, with a little ditty, our *Marseillaise*! It was a good day...

———

French sailors at Mudros. Their cap band suggests they are from the minesweeper *Caudan*.

(Ernest Brooks, Admiralty official photographer)

The French commissariat, Mudros.
In the distance, at the foot of the hill, right of centre, can
be seen the huts and tents of Evacuation Hospital no. 1.

(Ernest Brooks, Admiralty official photographer)

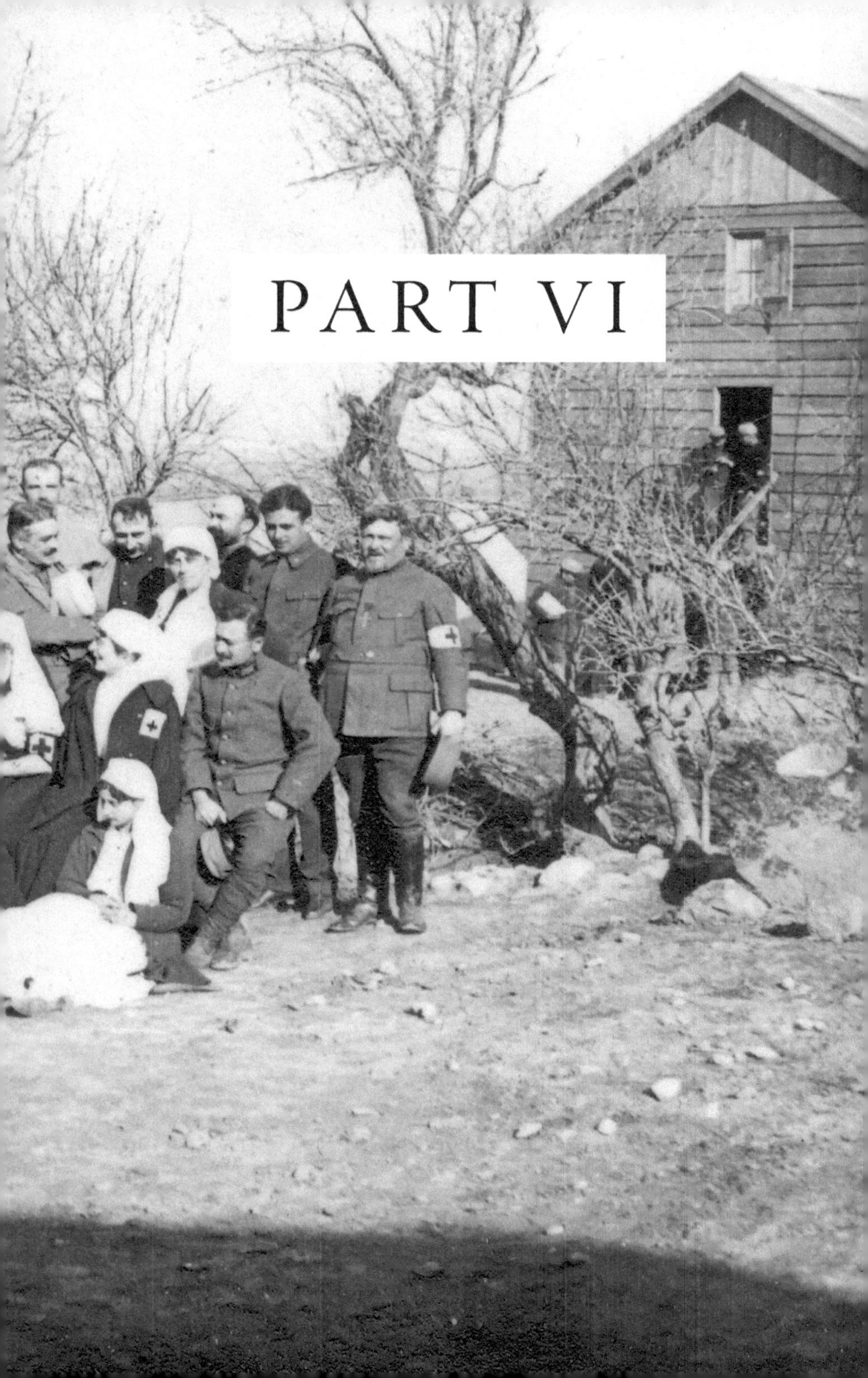
PART VI

One of a series of photographs that appear to
have been taken at Christmas-time.

(Elisabeth Jardin Collection)

I

Mudros, 1st January 1916.

And so... last night... the twelve strokes of midnight were rung out by twelve magnificent blasts from the sirens in the roadstead... All the ships had concerted the signal, and with surprising unity, they all howled into the night twelve times, twelve cries... twelve immense cries... It was not mournful; it was very grand. It was like a call to the heavens, or perhaps an act of suzerainty... Had we not the audacity to believe that it was *we* who regulated time...

The night was grave, solemn. There was no moon, and the stars shone, very dutiful and still. It was a new year... To make wishes? One hardly dared... Songs rose from the neighbouring camps, hymns that the British soldiers were singing to celebrate New Year's Day...

For our part, in our camp, we listened in silence. Amidst this ring of mountains, the sirens had resumed their songs, rising high, ever higher towards the heavens... The sea occasionally caught the trail of a flare... Then, since everything has an end, the songs ceased, the sirens fell silent, silence returned... and everyone reflected that it was now, indeed, the 1st of January, 1916...

We had a sweet surprise, a lovely surprise. About a dozen 'sailors of the Republic' turned up without warning. Under their arms, in very neat cases, were mandolins and guitars... They had come to serenade us, to thank us for coming so far to nurse their comrades...

We showed them into the room that serves as both our sitting room and dining room... As it happened, a pile of oranges remained on the table, with tiny Allied flags of very bright silk perched prominently right in their midst...

We seated them around the large table... And I assure you it was a charming sight. Their hands and faces, carefully scrubbed, shone as was fitting. The well-starched blue collars framed their beardless faces and brightened the group... With boyish awkwardness and boyish gestures, our sailors spread out their music; then, after a quick glance at the leader—a glance full of anxiety and timidity—they began, upon my word, to charm us in earnest.

It was simple, it was lovely... On this New Year's Day, in our wooden hut, so far from the soil of France, we felt nonetheless among family... We were moved, and we were happy... They played for a long time, their eyes utterly delighted by our approval. We served them tea... And these great strapping sailors, turned back into children, accepted the slices of cake we offered them with trembling hands. We no longer knew whether we ought to thank them for the serenade or for the sweet emotion they had stirred in us. It was so touching, this idea of coming here to tell us their thoughts in their own way... We raised our glasses in honour of France, and we drank to our mutual health. Then, just as they were about to leave, one of them drew himself up... He recited to us, in a hesitant and subdued voice, that famous verse by Mayol[1] which speaks of the Ladies of France...

———

1   Félix Mayol, a famous French café-concert singer of the Belle Époque.

(Elisabeth Jardin Collection)

178

Mudros, January 1916.

We have received a Christmas tree... It is the *Gaulois* that sends it to us... It comes straight from France... It arrives late, but what joy all the same... We formed a circle around the large crate, then, with quite motherly care, we lifted it out and removed, one by one, the innumerable ties that held it fast, to keep it safe...

It stretched its limbs with rapture beneath our north wind... The needles were quite fresh... And our fir tree straightened up, reborn to life...

We made a great fuss of this tree which came from France. Just think, on our island, it was the only one of its kind... It was a fine young fir from the Vosges, with a little of its native soil in its roots... We stroked it, we breathed in its scent, we looked upon it with respect... We gave it a fine place, well in view, and sturdy artillerymen dug the hole it was to inhabit. A ring of stones surrounds it, stones which have been whitewashed... From the top of the hill our fir tree strikes the eye... And all who come to our hospital have a look for it...

The men watch over its health, and on days of high wind, a shelter protects it. People sit all around it, too, to chat. 'Alsace Square' is one of the most popular spots.

You see, we find the joys of children in these far-off lands... We have the souls of children... This tree from France adds a gentle note...

And besides, the slightest vegetation brings back a little life... I still remember last August... I remember that makeshift garden which our patients had laid out... A prickly pear, and another plant picked up goodness knows where, whose leaves hung pitifully, withered to the very tips, were the only ornaments of this little circle... The men saved their water to give it a drink. And then they had placed fans all around to help the wind... It was so hot... The plants would be cooler that way... And the patients would remain there for hours, seated on benches we had fashioned, facing their garden which they gazed at constantly... It was a little greenery... It was, as I told you, a little life...

———

III

Mudros, January 1916.

I am going to tell you why, this evening, all the patients in our hospital—more than fifteen hundred of them—received a large distribution of oranges... I wager that you will be moved...

You know the hill opposite? Well, for two days now a regiment has been camped there. They have come straight from Sedd-el-Bahr. They have pitched their bivouacs. These have climbed all the way up the mountain, looking like a vast anthill... It is pretty, moreover, all those little tents which seem to hug the earth. Seeing them from here, one asks oneself how anyone can live underneath. But it appears there is room and one is not too badly off. All those mushrooms there, they shelter men, and sometimes, in the night, one spots the glow of a brazier slipping out from under the canvas.

Yesterday we went to take them tobacco. We filled the pipes ourselves, and it was a real pleasure to see their astonishment and their joy... Just think how many months they had been over there, and now women of France, in the white dress they venerate, had come in person to bid them welcome. It is true, too, that the youngest among us brought an exquisite grace to filling the old, seasoned pipes with her very fine fingers... We went from one to another, distributing all we had. Meanwhile, we chatted, and they were so moved that they stammered a little. Nevertheless, we knew what they had done, for this colonial regiment had been in the breach for several months.

But the most touching thing of all is what I am about to tell you...

You know that yesterday was Twelfth Night. One of our companions had the happy idea of getting the camp cook to bake a great *galette*—an enormous cake. And this cake was so broad that it could scarcely fit into the large basket, all ribboned with tricolour paper, in which it had been placed with such care.

Thus escorted, three of us set off at nightfall towards the camp where fires were being lit. That evening, the night was beautiful, the sky beginning to brighten, and Venus, the first to appear, watched over us with a benevolent eye.

We crossed the bare plain and began to climb the hill. During the day, we had attended to the men; now we had to think of the worthy NCOs, almost all of them regulars. We were shown the mess where they were dining. It was a tent just like the others, but to have more room they had dug deep into the earth, so that the canvas was nothing more than a roof. Some eight steps led down to it. Unannounced, we descended, the youngest holding with both hands the beautiful *galette* which still reigned so magnificently amidst the tricolour ribbons. It was simple and pretty, that white dress gliding underground...

The look on our men's faces at this apparition was a strange one, no doubt. They certainly thought it was something out of a fairy tale... Then, little by little, they understood... Well, by the light of their lantern, not a single one remained unmoved. There were real tears in the depths of their eyes... Men's tears in the eyes of children. They all rose, and at the few words of welcome and gentle comradeship they heard, they stood for some moments unable to reply...

Ah, these soldiers, what children they are! And how well they knew how to show their thanks...

At first, they all wanted to speak at once, and then, as that did not work very well, they turned in perfect unison towards their leader, the one who presided over the mess...

He spoke to us then of sweet and good things. He sought no fine phrases, only to say what was in his heart, what was in all their hearts. And then he went to fetch a fine tin mug on which had been engraved the silhouette of the *château d'Europe*, bearing the name Sedd-el-Bahr, and where a fine crescent and star recalled their fight against the Turks...

We had to drink their health in Samos wine, carefully retrieved from a soldier's pack. We drank to France first, and then to all the heroic soldiers of the Dardanelles... the dead as well as the living...

We passed the fine mug from hand to hand, and upon my word, never was a toast so simple, so sincere, and so devoid of artifice.

Our hearts were a little wrung. Yet how good it was to drink deep of this healthy, strong life. Those men looked at us with eyes so direct, so honest. Never had I felt, as I did that evening, the great protection surrounding us. Brothers could not have been kinder. What respect they showed in their slightest gestures. I assure you, you had to see it...

But that was not all, as you shall see...

When we had each drunk in our turn, they all exchanged a glance. Then the senior man drew from beneath his greatcoat a fine five-franc piece, put it at the bottom of the mug and passed it to his neighbour. They did the same, one after another, and before we understood, the mug was back in our hands. It was too small to hold all the silver coins. Some fell out: "For the patients of your hospital, ladies; we did not know how to tell you our thoughts, will you be our interpreters..." They insisted: "Please, let it be, it gives us such pleasure. If you knew the joy you have caused us... you see, now, we would fight with even better heart... Think that you have brought us a reflection of the families we left behind." He insisted. "You will buy them oranges..." They were moved to tears, the brave lads...

And one felt they spoke the truth... In their underground dwelling, on benches dug from the very earth, there was loneliness; and then too, it was still—always—the war... While here... Ah! Yes, how happy they were...

Now, as we made our way back, escorted by all the good wishes they had expressed, our souls felt at peace. It seemed that this was a blessed night, a night such as one would not often find again, and for having tasted it, one had to thank God...

In the sky, the stars had come out without mystery... They chased one another across the heavens, and we knew that in their turn they would go and tell the Divine what they had seen that evening... And we felt our souls as light as on days when one is content with the holy minutes God has deigned to grant... And the stars watched us still, gazing down, and that was for us a further sweetness...

———

Convalescent soldiers at Evacuation Hospital no. 1, Mudros.
The man on the left may have suffered frostbitten feet.
The man on the right sports the *Croix de Guerre*.
(*Elisabeth Jardin Collection*)

IV

Mudros, January 1916.

Oh yes, we were moved when we learned that Sedd-el-Bahr was to be evacuated once and for all, on the night of the 8th to the 9th. For several days, troops had been arriving from over there, full of spirit and bite, all ready to go and fight, provided it was not on that devil of a peninsula...

"We've seen enough of it, and for all the dust we ate there, we might as well go elsewhere... But all the same, had we been ordered to stay, we'd have fought on"...

Ah, these soldiers of France, they are all the same. Good, frank, simple, intelligent. And always a touch of mischief, a laughing word that serves sometimes to hide the surge of emotion making the heart beat faster... Ah, that French scepticism, how well we know it now... There are no finer, more honest lads than our soldiers... And knowing how to make the best of the smallest things, putting up with everything, grumbling as a matter of form, and then—ask them a service... The pride they feel in knowing themselves useful... And the gentleness they feel when shown they can do good, so much good! Ah, these overgrown children, what brave lads they were! ...

187

Indeed, by living in their midst, by sharing their life, one comes to know them better, to appreciate them more...

So, when we learned that each evening some of the men over there were being shipped out, and that finally only a handful remained, and that this handful—representing quite a few human lives—was to be taken off in turn in one single, final stroke... well... I will not hide it, we felt a chill, a deep chill in our hearts... It seemed a weight too heavy for us had settled upon our shoulders...

Rumour had it that the British would remain to the last to bear the brunt. We knew, too, that the British pier had been destroyed by a heavy sea and that only the French pier was in a fit state to be used... It was said further—and it was true, moreover—that the British had been forced to slaughter more than two thousand beasts, both horses and mules, and that they were jettisoning all the surplus with which they had encumbered themselves. Harnesses, motor-cars, everything went into the deep blue...

We French had embarked the greater part; the artillery remained, all ready to follow the rest... There were so many tales going about, true for the most part... Everyone returning from over there—and they were many—had their own yarns to tell... One saw the men filing past, groups of doctors coming to ask us for billets, streams of orderlies... Never had the island been so crowded...

Camps were being set up here and there... We met

old friends, we talked, and each of us, delighted no longer to be in that dreadful corner where there was nothing to do, began to relax... One felt a kind of wave of exhilaration preparing to rise...

At last the evening of the 8th arrived... Brrr... What apprehension! It was not that the elements were angry, but in the heart one felt a sensation like running ice...

Night had fallen, a true night of the Near East, filled with stars. The wind was blowing from the south, gently still... Two of my companions and I, arming ourselves with sticks and carrying lanterns, set off on the stroke of ten to climb to the very summit of the hill which overlooks our hospital. Despite the clarity of the sky, the night was dark, for the moon, still quite new, set early that evening... As we climbed, we saw it sinking. That embryo of a moon, that great red crescent—red in an almost unnatural way, fire-red—set in the night like a great drop of blood... I do not know why I found it so grim of aspect, with too much mystery about it. It struck straight for the horizon, drawing down with it stars that looked none too pleased...

High above, other stars were strolling at their ease despite the wind which was beginning to pant... The white dresses billowed, our capes flapped like wings and, struggling to keep our balance, we stumbled among the stones...

A great gust extinguished one of the lanterns. So we drew closer together to better sustain the struggle... Each step threatened to sweep the group away! We climbed, we climbed in the black night... And our ascent turned somewhat tragic in the great, oppressive silence. We cursed the wind which was going to jeopardise everything, and we suffered for all those who were over there... Oh, that walk at night, in the night of the 8th to the 9th... The stars might well have smiled sweetly upon us, but we paid them no heed. The wind alone preoccupied us... especially as it was blowing from the south...

A second lantern went out, struck full in the face by a sudden gust. Mine alone still shivered, jolted by our steps. In the end, it too fell silent, just like the others, so that we found ourselves at the summit of the mountain with our three dead lanterns... But what did it matter! The eye grows accustomed to any darkness. And then, my God, the stars up there were surely worth something... Besides, we were on a plateau with a stone wall right up against us...

It was half a sheepfold that had been thrown together right in the middle of the ridge... Stones piled one on top of the other to build a solid wall... It was one of those shelters of which there were so many on the island, used by Greek shepherds on days of high wind...

We faced towards Sedd-el-Bahr... So it was better to wait; if we were to see something, this was still the best place to be.

For I forgot to tell you... The object of our pilgrimage was to catch sight, if possible, of the great fire our troops were to light as they left... We had been told it was set for ten o'clock at night. Others, it is true, assured us it was for four o'clock in the morning...

The latter were right... We saw a great flash that seemed to slash the horizon. But the minutes passed and we saw nothing more... Only, the wind grew stronger and stronger, and the stars seemed more distant. Our great red crescent had slipped down to the sea; there was nothing left in the sky but a black backdrop where the golden pinpricks did not seem at all at ease...

The wind buffeted us with renewed force, snatched at our capes, froze us... We saw nothing, could guess at nothing... With miracles of dexterity, we crouched behind the stone wall and, after a thousand struggles, managed to relight one of the lanterns. The wind, each time we struck a match, snuffed out the golden flame as if with a hand...

Disappointed and downcast because of the wind which was blowing and which kept rising, rising still as though issuing from unknown depths, we started back... We stumbled often among the stones, but our minds were elsewhere and we cared nothing for it... Our soldiers, who were over there, what was to become of them? We did not feel very brave that night...

But when, towards two o'clock, I was woken with a start because my shutter was banging fit to break, I assure you I felt faint. Ah, how vile that wind was and how hard it struck... I threw my window wide open... It was indeed from the south that it came... It roared, it roared and it shook me to the core... In the roadstead, the lights of the ships seemed to be shifting... It was a veritable tempest!

How then were they to embark?... When the south wind blows at Cape Helles, there is no way to come alongside... With every swell, the waves swamp the meagre landing-stage. It is a frightful, ceaseless struggle... And one is helpless against it... Oh, if you knew how, that day, I suffered from my powerlessness... What rage filled my heart, and what pain too... Oh, that wind, how I hated it... And how I thought of all those who were over there... What an awakening it was...

But listen. The next day, I learned that the embarkation had been completed by the time the storm arose. British and French had abandoned the peninsula without loss of life... Only three wounded, and slightly at that...

At four o'clock, eye-witnesses who were standing out to sea on the ships saw the tremendous blaze lit by the Allies. They had left nothing to the Turks, who must have been baffled when, in the black night, they saw tearing through the darkness a sheaf of fire, said to reach two hundred metres in height.

It was, they say, a fairy scene. The shores of Europe and Asia appeared transfigured. They seemed bathed in red, illuminating the night.

———

VI

Mudros, January 1916.

They brought us Clairon... Clairon—you do not know him. Clairon is a regimental dog, the dog of the 54th Colonial... And as Clairon had no leave to take the road again, Clairon fell to our lot...

Ah! that Clairon, what a dog! He was all brown, and he could be witty, daring, impudent, like all those dogs who come from goodness knows where and whose pedigree is unknown. I wager Clairon never knew what a birth certificate was. Besides, would he have cared? I do not think so... He had a way of looking you straight in the eye and saying: "It is I, Clairon..." That was enough, was it not?... Oh, he knew very well what he wanted. So, whenever he heard men marching and caught sight of the ammunition boxes, it was quite a business holding him back. We had to find a rope almost as thick as a cable to put round his neck, and there, securely tethered, he would howl, turning eyes full of sorrow upon you...

Or else, when he had at last forgotten his regiment for a few minutes, he would jump up on your lap without leave, his paws caked with dirt. And if you could have seen his surprise when he was driven off... It is true that this did not happen very often...

193 "

For Clairon had become something of a child of the house, and one felt a tenderness and indulgence for this living creature who belonged, in a way, to everyone...

You should have seen Clairon the day of the review when French decorations were distributed to the British troops. Clairon, naturally, was of the party—our orderly had not the heart to leave him at home. Clairon pulled on his rope; the orderly followed. Clairon, heedless of discipline, slipped between the legs of the onlookers, and our orderly grew frantic. Finally, Clairon managed things so well that he arrived first...

He began to sniff the air in all directions as if to find his old comrades, then, doubtless disappointed, he sat down gravely and watched the Scottish regiment, with the pipers at its head, march past— regarding them almost with contempt: foreigners, to be sure!

Of course, that meant little to him...

Ah, that Clairon, how he would run through the hospital, despite orders to the contrary. But just try preventing him from seeking out a few men of the 54th Colonial who were being treated there... He would recognise them, wagging his tail, ears lowered, gazing with soft eyes in token of friendly greeting, then he would move on to others, continuing his rounds quite fearlessly.

But what I forgot to tell you is that Clairon has a soldier's service book. Almost a real book, with the regimental stamp as well... It was handed to us, with Clairon on the lead. And that booklet is well worth reading.

First CLAIRON, in large capital letters. Then: "Clairon, Involuntary Volunteer..." Under 'Service Record': he was at Valbonne, then from there, involuntarily, made the landing at Kum Kale; he shared the regiment's fortunes, involuntarily; went down to Sedd-el-Bahr... Involuntarily, always, he went through the whole campaign...

Under "Punishments," one reads:

"—Three days on the rope for nibbling a loaf without leave.

"—Four days on the rope for going out without leave.

"—Two days on the rope for failing to answer roll-call..."

And all this with the regimental stamp, as if to testify to its authenticity. And then, on the last page, are set down the names of all those who wish to keep in touch with Clairon and who take an interest in him. And I assure you, there are some who love Clairon...

We, for our part, have promised to bring Clairon back to France. And that is almost a point of honour...

————

On the waterfront at Mudros.

(Elisabeth Jardin Collection)

VI

Mudros, January 1916.

At four o'clock in the morning, drum rolls and bugle calls penetrated our hut. Rolls and calls that went on and on, seeming to take possession of the very air. The cold morning silence, which hushes everything, made this unaccustomed awakening the more piercing. The weary stars slipped from the heavens as if hastening to sleep. A pale chill rose from earth to sky...

At seven o'clock, when I went out, I learned the reason for this unaccustomed awakening. Our colonial regiment, encamped on the hill opposite, had departed in its turn. There were no more tents, no more joyful comings and goings. The site was bare. Only, marking the place where the tents had stood, great fires were burning. The colonials have their principles. Everything must be put back in order before their departure. So they had set fire to what they could not take with them; they were burning the straw and the refuse...

All these fires ranging in tiers upon the hill, while the sun scarce thought to show itself—all these fires shrouded in a great white mist burned alone, as if intent on fulfilling their task... They were innumerable, great and small; driven by a light wind, they all ran down towards the roadstead...

This departure left a great void. For several days we had grown used to seeing them... And from as far off as they could see us, they would snap into step to render their salute smartly; they were good lads, brave lads, ready to lay down their lives. And while the sun in its turn came to resume its task, the fires, one by one, grew dim. We watched them pale, sink down, then die, as if, with the coming of day, their life were departing too.

———

(Elisabeth Jardin Collection)

198

VII

Mudros, January 1916.

We must leave... We must bid farewell to things and to people. We must leave... The months have passed, kind and sweet. We were all one family, one heart with one desire... We all thought of France above all. It was a very wholesome, very lofty atmosphere, an atmosphere that left pettiness and ugliness so far behind... It was a great brotherhood, a touching solidarity... And it was to all this that we had to bid farewell...

———

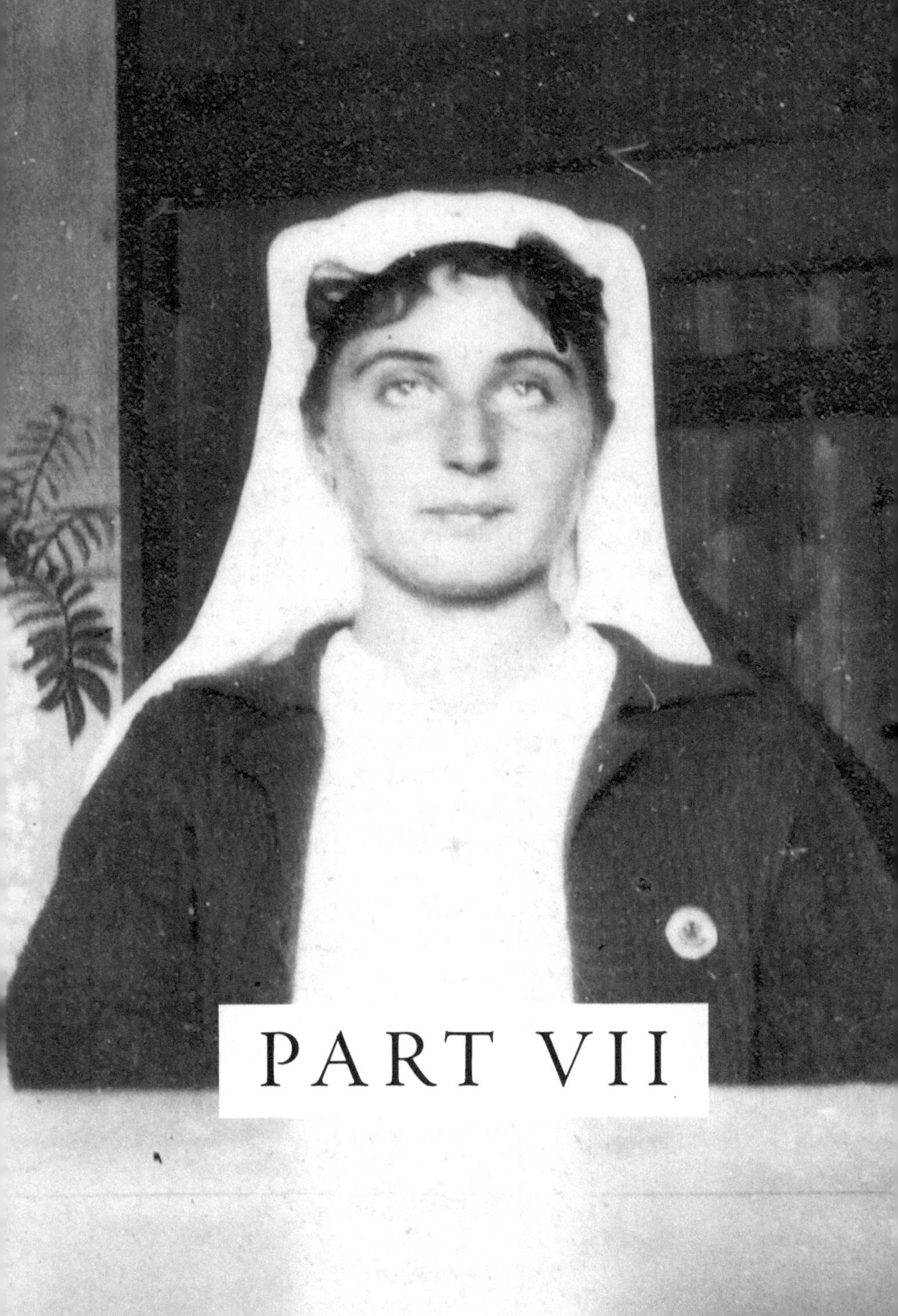

PART VII

The British Admiral C—[1] came on board to bid me farewell, as did the General commanding the Mudros base and his staff officers. Most of the doctors from our hospital—a group dwindling day by day, for the dispersal has begun—and my companions are here. The day before, I had received a visit from the British general and the British colonels C—, B— F.M.C—, old friends of my family rediscovered here.

Admiral C—, remembering that we are in wartime, had sent me a few days before, with charming simplicity, some 'comforts' for the voyage: a tin of biscuits, a tin of cocoa, and a tin of milk! He did not know I was to travel on so fine a ship, and so he apologised...

It is to all this sweet simplicity of our daily life that I must say farewell, to our hospital which I glimpse from afar, to all those huts that belonged to me, those five huts which made up the 2nd Fever Ward. I voiced my regrets and my good wishes to every patient... We shook hands, and for the last time I looked over, one by one, the smallest nooks and crannies of our hospital... Life there was hard

1 Rear-Admiral Arthur Christian, Royal Navy.

at the beginning, but improvements came, and it had become magnificent! Even narcissi had seen the light of day there...

I leave a large part of myself there, but I take much away. It is true, this camp life was sweet to me. I drew new energy from it. A great inner peace and a great serenity of soul are the riches I gathered there... Through having seen suffering and death, one knows better, surely... One knows better that life is but a passage, one knows better what is within oneself and what one has to give... One learns to know oneself. Each of us can do good, for each of us has something of good within... It is enough to strike the spark, enough to look for it in others; and in seeking it, one has already gained much...

I recall the long conversations we had in the evening with my admirable friend, our Matron, and another of my companions, another very beautiful soul... I tried to follow them... With a common desire, we sought the goal to be attained. We wished to make ourselves better by forgetting ourselves... We would turn the pages of the Holy Bible or the Holy Gospels, reading them aloud, discussing them... And then, above all, I often threw my window wide open to see my stars, for I still maintained that the supreme beauty of nature was the fairest manifestation of God... And that, with nights such as those, in an atmosphere such as we breathed, one could not but be good...

What great dreams of peace and kindness we dreamt on those evenings...

But we must leave, we must say farewell to all those who strove to spare us the difficulties—illusory ones, at that—which we might have encountered. We must thank everyone, for high and low alike spared no effort... We were all one family on our island...

But it was not until the 7th of February that I finally left Mudros, aboard the auxiliary cruiser *Provence II*. We weighed anchor early in the morning after hoisting the regulation signals. And then we got under way slowly, gliding gently over the grey water, until we were beyond the mine barrage... But once outside the enclosure, we steamed ahead at full speed...

It was a sad morning, with a cold mist that failed to drive away the winter wind... We had on board some three hundred Turkish prisoners, nearly a hundred and fifty officers on leave, and a number of soldiers... But all this disappeared in the great floating city that was our *Provence*. She had been clad in a magnificent dark grey to blend her with the sea...

From the bridge, one looked down upon almost the entire ship, and from the bridge one could see far, very far... The look-outs were relieved every hour. The fine 14-cm guns, smooth in their coats of paint, stretched out... The 47-mms, sharper, more brazen, pointed to the sky... Everything went well;

the Captain was satisfied. As for the men, they rejoiced to be returning home.

At luncheon, I was placed at the senior officers' table... And naturally, we spoke of the war... Of what else could we have spoken?... And as most of them had been at Gallipoli, they recalled shared memories... The arrival at Kum Kale... The landing parties who had done their duty so magnificently, those companies that had returned with but twenty-five men...[1] the piles of dead and wounded, the heroism of the troops... and the battery which, perched right at the top of Kereves Dere, threatened Achi Baba. And I can still hear Major X— saying to me:

"You know, when you've been through that whole campaign, as I have, you've truly earned your leave."

Another added, with a gentle dream in his eyes and a tinge of anxiety:

"And then... it will be good to be back with the family... I wonder if I shall have the luck to see my big lad of a son again... He was in Champagne..."[2]

Yet another said to me:

"You saw Morto Bay, you saw all those graves there... well, I had one of my lieutenants killed beside me, right there; he was cut in two by a shell..."

1 A company in the French army had a maximum strength of about 250 men.

2 The French conducted major offensives against German lines in the Champagne region between December 1914 and November 1915, resulting in massive casualties for both sides.

Then he went on:

"The hardest thing for me in the evacuation of Gallipoli—I know well that nothing could be done about it—was leaving behind all our dead..."

After a pause, he continued:

"I believe, for my part, that the Turks will respect them... They respect the wounded and the dead... They never fired on the barges carrying the wounded away..."

Then everyone spoke at once... Someone called out:

"You know what they call us? The duffers of the Dardanelles." [1]

We spoke also of several of their comrades who, stricken with typhoid, had come to die in our hospital:

"Poor old X—, what a good fellow he was. Always in high spirits, always ready to make for the most dangerous spot... Did he suffer much?"

At our table sat two Serbian officers and a British staff officer... The former, pensive and sad, listened without a word... They listened as one listens in a dream... They listened with drawn faces... And I felt pity, we all felt pity, for that great silent sorrow...

We spoke then of Serbia, in hushed tones, as if speaking of the dead. We spoke of the retreat. We spoke of the Serbian soldiers dying at Vido and of the great barge that, every morning, bore all the corpses out to sea to commit them to the deep...

---

1 *Les ballots des Dardanelles.*

One of the Serbian officers uttered:

"How could it be otherwise? We went not days, but weeks without eating... My own family is still back there... Do they have bread? Have they been killed? I do not know..."

We spoke further of humanity, of the great lesson of the events we were living through... We were so small... One had to be kind... And yet we kept such great faith...

We philosophised for a long time... Then photographs were spread out. The slightest anecdotes connected to them were told to me, the slightest words repeated. There was the photograph of the children, of the mother... Photographs that had been feverishly clutched in the large wallet kept against the heart...

We lived a quiet life on board, for the *Provence* remained very steady... One evening, we went down almost to the bottom of the hold, wandering long in dark corridors to hear a concert improvised by soldiers and sailors. A Leading Seaman did the honours of their mess. And there, crowded together—officers, ships' boys, men of all ranks and ratings—we spent exquisite, happy hours. A soldier played for us on a mandolin of his own making, fashioned under the shells of Gallipoli, old tunes charming and sweet. A Provençal recited, like a true artist, monologues from his home, and the beautiful tongue of Provence resounded magnificently. Then, armed with a bottle into which he had slipped two spoons, he accompanied the band…

Then, of a morning, we would go and say hello to the foals born with the Expeditionary Corps. Some had first seen the light at Mudros, others at Sedd-el-Bahr… such as 'Marguerite,' a plucky little creature, not always easy to handle, who had been born on the 12th of June…

We sometimes spoke of the submarines we knew to be in the vicinity… We were a fine prize… But no one was alarmed… We were past that point… And besides, we would not have been the only ones to die for the country… Only, one day, one of those who had done his duty most magnificently since the start of hostilities let fall these words, as if answering a dream:

"Ah! No… to come out of the Dardanelles alive only to die stupidly like this, no… I can stomach the return trip, but at least let a man go and kiss his wife and little ones…"

Upon the sea, our wake traced magnificent zigzags. And, as the Captain was fond of saying:

"We are staggering like a drunkard…"

We steamed at eighteen knots, night and day. At night, I went up to the bridge, after reaching—not without difficulty, given the gloom that reigned within our *Provence*—the staircase leading to it… The night was black, thickened by banking clouds…

One could not see two metres ahead. The great roar of the sea we were cleaving struck the ship's flanks. That night, the sea no longer had its usual voice. It was magnificent and terrible, this mad dash through the tempestuous air. The howling of the engines, sustained by the rumble of the water we were crushing, made up one immense and unsuspected whole. The officers of the watch and the look-outs strained their eyes, trying to pierce the darkness. And Captain Vesco, ever faithful to his post, where he could be found at any hour of the day or night, also probed the gloom with a stern eye. Wrapped in his goatskin coat, he paced the bridge, giving an order, keeping a close watch on the vessel's course.

"You can see nothing... absolutely nothing... We are making eighteen knots..."

The enormous bulk of the *Provence* merged into the night...

Two days later, we hit a storm. The sea came aboard at will, and each time we buried our bows in a wave, a vast white sheet was flung up, covering the whole of the foredeck. The guns pivoted under the impact... Our fine 14-centimetre guns streamed with water. Then the Captain ordered speed reduced:

"I must have those children of mine looked after a little... I may have need of them... To-day, this weather—wretched weather for submarines..."

He said this gazing at the sea... He spoke in the manner of sailors, who seem always to be addressing the open ocean, and answering a question that they alone have heard...

Captain Vesco never left his post, and I wonder, during the four days the crossing lasted, how he endured it. We made immense detours to throw the enemy off the scent... Every evening, I went up there; many of us gathered, for we were all one family, for everyone considered me absolutely one of their own, and try as I might to fade into the background... I too had been on campaign... Oh, there was no resisting it... We said "us"... We were a solid block, the "Orient" block, of which we spoke ceaselessly... For we spoke of all the comrades left over there, the dead as well as the living. We almost never spoke of ourselves... We laughed with good, moved laughter at the memories we had amassed... And it was healthy... and it was young... And it was heartening... And it was the very essence of France...

You see, all those who have not lived that life, I pity them infinitely... One cannot know the impetus it gives to the soul... It is a door thrown open...

The *Provence* was sunk on her return voyage; they had been lying in wait for her for a long time... Captain Vesco died, heroically and simply, just as simply as he had done his duty until then... Knowing him, his end does not surprise... That "Farewell, my children" which he called out to the survivors, that farewell was him. He remained clinging to his bridge, supervising everything, to the very end... Refusing to be saved when he might have been... Yet, he was happy. He had a wife he adored, two children of whom he was proud... But his country came first...

With him died several of those I had known on board, for many were sailing on the *Provence* again... May they have found in France the joys they deserved, and may they carry with them all our immense gratitude and all our sorrowful respect... And glory to our sailors and to our soldiers... Glory to them... always.

* * *

BOOK TWO

*A casualty clearing station
at the Dardanelles*

—

EVACUATION HOSPITAL NO. 1, MUDROS

(APRIL 1915 – FEBRUARY 1916)

—

Elisabeth Jardin

Docteur Elisabeth FABRE

DE LA FACULTÉ DE MÉDECINE DE PARIS
EXTERNE DES HOPITAUX
EX AIDE-MAJOR A LA MISSION
ANTIPALUDIQUE DE L'ARMÉE D'ORIENT
DÉCORÉE DE LA CROIX DE GUERRE

Un Hôpital d'Évacuation aux Dardanelles

L'H. O. E. 1 DE MOUDROS

(AVRIL 1915 - FÉVRIER 1916)

PARIS

JOUVE & Cie, ÉDITEURS

15, Rue Racine, 15

1920

The work that forms the centrepiece of this second book is a doctoral thesis. Submitted in 1920 to the Faculty of Medicine of Paris under the married name Elisabeth Fabre, its full title—*A casualty clearing station at the Dardanelles: Evacuation Hospital no. 1, Mudros (April 1915–February 1916)*—announces a clinical purpose far removed from the lyrical memoir of Jeanne Antelme. Where she gave us the campaign in glimpses, Jardin's thesis gives us the hospital, its organisation, its case-loads and its pathologies. Her credentials—hospital extern; former assistant medical officer of the antimalarial mission, Army of the Orient; *Croix de Guerre*—speak of a career forged in the wards she describes.

Elisabeth Jardin was born into a military family. Her father, Colonel Armand Marie Jardin, was an infantry officer who rose through the ranks from Saint-Cyr to a professorship at the *École Supérieure de Guerre*—the staff college that trained the senior officers of the French army. A veteran of the Tonkin campaign, he was decorated with the *Légion d'honneur* and the Japanese Order of the Sacred Treasure before his death in Paris in 1909, aged 55. Her grandfather, Édélestan Jardin, had served as an inspector in the Marine, a naturalist of some distinction who collected botanical specimens across the globe and presided over the *Société de Géographie* in Brest. Elisabeth's mother, Henriette Maillin, came from Quimper in Finistère; it was there that her parents had married, and there that Colonel Jardin was buried.

Of the colonel's seven children, at least three served in the Great War. Henri Jardin was a colonial administrator in Indochina. Elisabeth and her sister Marguerite both deployed as nurses—Elisabeth to Mudros and Salonika, Marguerite to the field ambulance at Veria in Macedonia, where she was awarded the Russian Order of Merit.

Unlike Jeanne Antelme, an aristocratic volunteer, Jardin was a medical student, an *étudiante en médecine*, providing her with a clinical grounding that distinguishes both her service and her later writing. In July 1915, she boarded the hospital ship *Duguay Trouin* with the first echelon of French nurses for the Dardanelles, led by Matron *Mademoiselle* Oberkampf. They disembarked at Mudros to staff Evacuation Hospital no. 1, the sprawling facility on the hillside south of the town that would grow to become the largest hospital on the island. This timeline places Jardin among the very first female nurses on Lemnos, arriving weeks before the second French echelon and predating the first Australian sisters, who landed on 5 August.

On 12 April 1917, Jardin was mentioned in despatches and awarded the *Croix de Guerre*.

The citation described her as a 'nurse of tireless activity and unfailing devotion' who had proved 'a precious collaborator to the treating physician of her ward.' By this time, Evacuation Hospital no. 1 had moved to Zeitenlik on the outskirts of Salonika, where it was reconstituted as Auxiliary Hospital no. 7.

After the war, Jardin completed her medical degree and married Eugène Fabre in 1919. At her death in May 1929, she was recorded as *Madame le Docteur* Elisabeth Jardin—a reversion to her maiden name that echoes Antelme's own return to hers after her divorce from Noblemaire.

* * *

This volume features many photographs drawn from an album compiled by Jardin during her service. The photos surfaced on the online marketplace eBay in 2021. Although the integrity of the original album was compromised when the collection was broken up and sold piece by piece, a far greater tragedy had been averted: the seller revealed that the album had originally been rescued from a rubbish bin in Paris. The translator, alongside Bill Sellars, was able to purchase a significant number of these photographs, reuniting Jardin's images with written accounts of the hospital.

Jardin's album was divided into two sections. The first documented her time on Lemnos; the second captured the French medical mission on the island of Vido, off Corfu, in early 1916, where the lens caught the harrowing aftermath of the Serbian army's great retreat across the Albanian mountains. French medical teams battled a catastrophic typhus epidemic among the emaciated survivors on this 'island of death'; over five thousand bodies were weighted with rocks and consigned to the sea.

Among the items in the collection is a postcard written on New Year's Eve 1914 by a young sailor, Henri Cazenabe, aboard the battleship *Suffren* at Toulon. He announces to his sisters that he is leaving the next morning for the Dardanelles to 'bombard the Turks,' and encloses a photograph of himself and his companions, taken at night, 'a little tipsy.' He was eighteen years old. The *Suffren* survived the Dardanelles campaign, but on 25 November 1916, en route to Lorient for refitting, she was torpedoed by a German submarine off Lisbon. A magazine exploded and the ship sank within seconds; all 648 crew were lost. Among Jardin's photographs is one showing her on a launch in Mudros harbour, heading out to visit the *Suffren*—a reminder that the sailors of the fleet and the nurses of the hospital shared the same waters and the same war.

B.dB.

CONTENTS

(Elisabeth Jardin Collection)

Elisabeth Jardin.
The cross and letters 'S.B.M.' on her pinafore identify
her as a nurse with the *Société de secours aux blessés
militaires*, the oldest branch of the French Red Cross.

(Elisabeth Jardin Collection)

DEDICATION

To the memory of my father, Colonel Jardin, Professor at the *École de Guerre*;[1] to my husband; to my mother; to my brother; to my sister; to my family and friends.

To my teachers in the Paris hospitals—Messrs Rist (placement, 1911–1912), Barth (placement, 1911–1912), Renon (placement, 1911–1912), Jalaguier (extern,[2] 1913), Savariaud (extern, 1913), Guillemot (extern, 1914), Bar (extern, 1915).

To Professor Carnot—I am grateful to him for the honour of his presiding over this thesis.

The name of Mudros remains for me inseparable from many warm memories of friendship with my companions in the nursing team of the *Société de secours aux blessés militaires*.

The doctors of the hospital showed every proof of true comradeship, and I thank them for the documents they so generously provided, which were of great assistance in this work.

---

1 *École de Guerre*: the French Army's senior staff college in Paris, responsible for training officers for high command.

2 Extern *(externe des hôpitaux de Paris)*: a competitive junior hospital appointment in the French system, ranking above a general clinical placement *(stage)* and below an *interne*. It has no precise British equivalent.

From the senior medical officer of Evacuation Hospital no. 1, a warm and collegial welcome was always forthcoming, along with invaluable support in gathering the material for this thesis. May he accept this expression of my deepest gratitude and most sincere thanks.

British naval officers speak with three French nurses of Evacuation Hospital no. 1.

The two nurses, from left: Elisabeth Jardin, and probably Jeanne Antelme, with notebook.

(Ernest Brooks, Admiralty official photographer)

PREFACE

In July 1915, the *Croix-Rouge (Société de secours aux blessés militaires)* decided to send a team of nurses to the Dardanelles expeditionary force. The team consisted of six nurses under the direction of a head nurse.

Having had the honour of serving in this team, which was attached to Evacuation Hospital no. 1 at Mudros, it seemed to me worth describing how this hospital operated—an institution whose life unfolded amid somewhat unusual and often difficult conditions—and studying the diseases that had to be treated there.

May I be permitted, as I begin this review through so many memories, to offer my modest tribute to the unceasing and devoted efforts of all those who, in that hospital, far from the mother country—which many had never before left—under a trying climate during the long summer months, and in a land forsaken by nature, applied their intelligence and their hearts to accomplishing good medical work.

The island of Lemnos, with its volcanic and rocky soil, like most of the islands of the Aegean Sea, supports only a precarious vegetation, confined to the valleys and even there limited to the brief duration of spring. Trees are almost entirely absent, as in many lands that were subject to Turkish rule. The tormented landscape where antiquity placed the forges of Hephaestus may offer the poet impressions—rendered by more than one writer in delicate terms—but its resources were ill-suited to the establishment of a hospital. No trees, scarce water, high temperatures during the summer, a sparse population of fishermen or shepherds reduced to a wretched existence, primitive and barely habitable houses—in short, nothing existed on the island that could make the stay comfortable for the sick and wounded.

There was, however, no question of choosing another site for the hospital.

The Dardanelles expeditionary corps depended for its survival on a substantial fleet—naval, merchant, and hospital. This fleet had found in Mudros harbour a natural and unequalled shelter against the sea and against submarines. The naval base had been established at the head of the bay, where the small village of Mudros offered, ready-made if rudimentary, the first jetty for disembarking troops and matériel. It was as close as possible to this base, the source of all supplies, that Evacuation Hospital no. 1 was to be sited. A rudimentary casualty clearing station to begin with, but one whose development proceeded with such rapidity that within a few weeks it had become a fully-fledged hospital *(hôpital temporaire),*[1] its patient population exceeding 1,500 sick and wounded.

In this work, I propose to review the establishment of this hospital, then its operation, and finally—in greater detail—the various departments of this bustling institution, together with a study of the conditions treated there.

———

1 In the French army medical service, the *hôpital d'évacuation* (or *hôpital d'origine d'étapes*, HOE) was a forward medical formation in the army zone, typically near a railhead, designed for triage, short-term stabilisation, and the rapid onward evacuation of casualties. It was not intended for prolonged treatment. The British equivalent was a casualty clearing station. The *hôpital temporaire*, by contrast, was a rear-area facility established in requisitioned civilian premises—hotels, schools, châteaux—for sustained care and convalescence, well removed from the front. That the evacuation hospital at Mudros evolved into a *hôpital temporaire* reflects the exceptional circumstances of the Dardanelles campaign, where difficulties of evacuation by sea forced a transit facility to take on a role far beyond its intended purpose.

CHAPTER I

Establishment of Evacuation Hospital no. 1

We shall recall a few articles from the Regulations on the Medical Service in the Field that define the functions of evacuation hospitals.

"The evacuation hospital is the essential unit of the rear-echelon medical service: It plays a significant role in the conservation of manpower by preventing the evacuation to the interior of a large number of sick or lightly wounded men who are capable of returning to duty at short notice, and whom it is therefore important to keep close to the army. It is at the evacuation hospital that all convoys or transport columns of the sick and wounded arrive, regardless of their origin. It provides temporary hospitalisation for those who cannot be transported further, and ensures the rapid evacuation of the others to hospitals or hospital centres in the rear, or towards the interior.

"The evacuation hospital is organised so as to be able to operate, if necessary, as two separate sections, each capable of treating 100 patients.

"Provisioning is ensured by drawing on the supplies of medical formations, or from the exploitation of local resources.

"The dietary regime in evacuation hospitals

comprises food and drink, and is prescribed as either a full ration or a half-ration.

"The nature and quantity of allocations are set by a special schedule given in Notice 5 of the Regulations on the Medical Service in the Field.

"The dietary regime set out above is that given to lightly sick and wounded who are not to be evacuated."

Let me say at once that the question of provisioning at Evacuation Hospital no. 1 was settled on a basis quite different from that laid down in the regulations. This initiative was driven by pressing necessities, foremost among which was always the welfare of the patients. We shall examine this point in greater detail in the detailed study of the hospital's departments.

Beyond its normal functions—namely, first, the evacuation of sick and wounded who could not be returned to duty, and second, the treatment of lightly sick and wounded likely to resume service within a short time—Evacuation Hospital no. 1 was compelled, from the very start of operations at the Dardanelles, to retain a considerable number of patients whom the difficulties of evacuation by sea, and the nature of the conditions from which they suffered (various contagious diseases, typhoid fever, dysentery, etc.), obliged it to treat on the spot.

Thus, within a few weeks, this unit took on an importance far exceeding the scope of an evacuation hospital, and in practice became a *hôpital temporaire*.

History

Evacuation Hospital no. 1 left Marseille on 3 March 1915, aboard the *Magellan*. After a four-day stop at Bizerta, it arrived in Mudros harbour on 15 March. Unloading of equipment began on 24 March. On the following day, the hospital took possession of the site reserved for it, approximately one kilometre from the naval base.

This site was admirably chosen. At the foot of a high hillside, occupying its lower slopes, the hospital rose in tiers before a magnificent panorama. The great anchorage spread itself in the foreground, dominated by the village of Mudros, all white on a slight promontory. A fleet such as those waters had never before carried gave it constant life—by day the ceaseless movement of its vessels, by night a thousand lights. Beyond, a chain of mountains traced its summits in harmonious lines, crowned on the clearest days by the bold cone of Mount Athos, so sharp despite its distance that it seemed part of the island itself. The purest light bathed this landscape, and the clear air wrapped it at every hour of day and night, passing through colours in endless variation. And each evening brought to those far from home a fresh emotion as the setting sun set the island and its roadstead ablaze; the poet longed then for the brushes of her painter friend.

* * *

CHAPTER I

Evenings of Lemnos! Some are golden as ripe fruit.
There floats within them an impalpable light,
That light unseen, wherein the Gods walk,
Woven perhaps of dust that
Dances from the very waves to heaven:
An evening fit for Pythagoras, quickened by numbers
Which do not outrun the Stars—for they are
Silent beyond the horizon's edge, and
Rise through the blue air without the prop of shadow.
Those Elysian evenings are made for your colours—
Those tender evenings, those tawny evenings,
*   those evenings of rose;*
They are made for the colours upon your brushes,
Those evenings whose light gold becomes the flesh of things.
Evenings for Greek temples and for sages of quiet wisdom.

Other evenings are like songs from the Iliad.
Invisible battles strike their lightnings through them;
Above, the warriors in the unseen ether.
The Gods still contend for ancient Troy;
Their great and splendid robes are mingled together,
They shift their colours from deep crimson to fresher hues,
Their spears make a field of wheat across one corner of
*   the sky,*
And suddenly the valleys are thick with their arrows.

The evenings when Heroes stand before the Altars
Those evenings of crimson are evenings for immortals.

The evenings that belong to me are strange and sad,
Cut from pale emerald and amethyst,
Nostalgic, fever-marked, uncommon;
They have held fast all the ardour of the summers,
And some low fitful wind carries to them
The treacherous soul of this dead harbour.
Their grey-green depths are mirrors like those
Of Magicians, wherein suns perish;
They shift at the mad whim of the clouds,
Coloured by fires that herald storms,
As shadow moves and covers in its turn
Those corners of earth made to hold the last of
*   the day.*

Such evenings are mine—evenings already half-night,
Saturnine, laden with cold colours,
Exchanging at the threshold of Mystery their vessels
With the Night, which alone can fill them with tears.
Hecate glides then as though through a dawn;
At the foot of the east lies a lake of light:
And there I shall find—with a heart consumed by
*   everything—*
Neither peace nor rest.

*   *   *

At the outset the hospital consisted of four *tentes tortoises*,[1] one of them reserved for infectious cases, capable in all of accommodating 200 stretchers, half of them on trestle supports. There were also five *marabout* tents: for the chief medical officer, for the administrator's office, for the duty medical officer, for the dispensary, and for equipment stores.

The troops of the expeditionary corps, which arrived at Mudros between 10 and 20 March 1915, were re-embarked on 25 March and sent to Alexandria. They returned to Mudros on 24 April, and on the following night departed for the Dardanelles.

During this period, from 24 March to 25 April, Evacuation Hospital no. 1 received few patients. These came from passing ships or from the small number of services left at the naval base. This lull was put to use on various works: construction of a road linking the hospital to the base, and the tapping of water from a spring near the hospital.

1 *Tente tortoise*: a canvas field tent adopted by the French military from a British design, so named for its broad, rounded profile which offered high stability in the wind. The French medical service deployed it in two distinct configurations. The standalone field hospital ward tent was large, capable of housing up to fifty stretchers, half of which could be raised on trestle supports *(supports-brancards)*. A smaller, highly mobile version was structurally integrated with a medical supply wagon *(fourgon d'ambulance)*; its canvas was unrolled and pitched directly around the vehicle to serve as a rapid-deployment surgical operating room.

The opportunity was also taken to improve the initial set-up, which had been carried out in haste. A wooden operating theatre was built, then kitchens; beds were fashioned from wood and rope; new *marabout* tents were erected; the ground was levelled; and small stone paths were laid out across the hospital site. From 26 April, the date on which operations began, until June, the hospital functioned with an average of five to six hundred sick and wounded. The most seriously affected were placed under *tentes tortoises*, lying on makeshift beds or on stretchers. The others were under *marabout* tents, lying on the ground on mattresses stuffed with seaweed or on hay.

This period was the most trying for the patients, but everyone made efforts to improve their situation. The hospital staff—medical officers and nurses alike—gave of themselves unstintingly. They also found a valuable source of assistance in the naval squadron lying at anchor in the roadstead. The navy placed beds, tables, and tarred-canvas baths at the patients' disposal, which proved most useful while awaiting the arrival of zinc bathtubs; it also provided them with comforts that would otherwise have been unobtainable: champagne, jam, fruit, and the like.

Towards the end of June, the number of patients increased considerably, soon reaching approximately 1,500. Construction of wooden huts then began. These were of various types—Bessonneau, Favaron,

Adrian, Egyptian huts[1]—and largely replaced the tents. By July the hospital had reached 1,600 patients, a figure it was not to exceed thereafter. In April the bacteriology laboratory had been established within the hospital, and a radiography laboratory followed. The engine installed for the latter could also provide electric lighting throughout the hospital. After the makeshift camp of those first months, this was comfort and luxury indeed!

At the same time, the hospital's capacity could be reduced to 1,000 beds, allowing greater ease across all departments. This measure followed the reorganisation of the expeditionary corps. At the end of September, a division was drawn from its strength and sent to Salonika, where the *Armée d'Orient* was being formed.

Around October, a trial stone-built structure was erected for the seriously ill patients in the infectious diseases service. Although successful, the experiment was not followed up, as the hospital was dissolved two months later.

A shower block was also constructed.

The rudimentary kitchen gave way to a substantial stone building. A cement water reservoir was

1 Bessonneau, Favaron, and Adrian were named types of prefabricated military hut, each associated with a manufacturer or designer. *Égyptiennes* does not appear in contemporary catalogues of standard types and probably refers to huts shipped from the French base in Egypt rather than to a proprietary design.

installed, in which water sterilisation could be carried out more methodically.

Finally, thanks to abundant supplies from the French and British Red Cross societies, patients were able to have linen in unlimited quantity.

Thus was Evacuation Hospital no. 1 built up and sustained—continually reorganised and constantly improved—operating at Mudros until February 1916. At that date, its essential elements were transferred to Salonika, where it became a very large *hôpital temporaire*.

———

Evacuation Hospital no. 1 with Mudros town in the distance.
The two white marquees are likely *tentes tortoises*.
(*Lieut. Homolle, Ministère de la Culture / APOR051214*)

CHAPTER II

General operations of Evacuation Hospital no. 1

The island location of our hospital at Mudros gave its operations a character distinct from that of an evacuation hospital on the mainland. As we have seen, from a very early stage it assumed the functions of a true *hôpital temporaire*. I shall now briefly describe the stages through which sick and wounded passed in reaching the hospital, and then how they were discharged from it.

Mudros is situated approximately 80 kilometres from Cape Helles. Using large transport vessels, this distance amounted to barely a few hours' crossing. But the difficulties and dangers presented by the approaches to the Gallipoli Peninsula, where fighting took place very close to the shore, soon made it necessary to risk only small craft in those waters, as they were less vulnerable both to enemy fire and to submarine attack.

Only the hospital ships were permitted to lie at some distance from the shore. By day they were identified by their white hulls bearing a broad green band running from stern to bow, and by night by a green row of electric lamps extending their entire length. This dual marking was compulsory for the navigation of these floating hospitals. They received only the severely wounded, who could be treated and operated on board as quickly as possible; the ships then ensured their evacuation to France or to Algeria.

The less serious cases remained in the field hospitals established on the peninsula.

There thus remained all the wounded with moderate or minor injuries, and the sick in varying degrees of severity. It was to Mudros that they were brought back, embarked on tugs returning from Cape Helles after completing resupply runs. The crossing lasted twelve hours on average and was never particularly comfortable on these vessels, where the patients occupied mainly the deck— the only space large enough. In bad weather, the crossing could be very harsh.

A small hospital ship, the *Saint François d'Assise*,[1] did handle the evacuation of a certain number of patients to Mudros, but its capacity was limited and it could not manage the whole work alone.

As soon as these vessels had landed their patients at Mudros—and where possible, as soon as their

1 The *Saint François d'Assise* was a three-masted steam-schooner of some 556 tons, launched at Nantes in 1901 as a floating hospital for the French cod-fishing fleets on the Grand Banks of Newfoundland and off Iceland. Operated by the *Société des Œuvres de Mer*, she carried a surgical ward, pharmacy, and around 20 hospital beds. During the First World War, she was requisitioned by the French Navy.

arrival in the roadstead was signalled—the naval base would notify the hospital. There, a medical officer was detailed each day for what was known as landing duty. He made his way as promptly as possible to the disembarkation point, accompanied by a party of orderlies and *araba* carts,[1] or later, by motor ambulances as well.

On the patients' arrival at the hospital, a special bell alerted all the medical officers, each of whom was required to report to his own ward. The duty medical officer went to the admissions office, where, at the same time as the incoming patients were registered, he assigned them to the various wards. This arrangement spared the patients excessively long waits, as they generally arrived in the evening or at night. They appreciated it especially when the bad season set in, and when, soaked and exhausted by a difficult and uncomfortable crossing, they had the satisfaction of finding, the moment they entered, hot drinks, shelter, and a bed.

The patients' stay in hospital could end, depending on their condition, in one of the following ways:

1. For those who had recovered, by discharge to the Mudros depot, with a view to their return to their unit;

---

1 *Araba*: a light two-wheeled cart drawn by horse or ox, common across North Africa and the Eastern Mediterranean. Familiar to French military personnel from colonial service in Algeria, the term was carried into use in other theatres.

2. For those who could return to active service,
    but only after a period of rest, by discharge to
    the convalescent depot established on the island
    near the hospital;

3. For the severely wounded and long-term
    convalescents, by evacuation on a hospital ship.

Evacuation typically involved large numbers, as
patients exhausted by the climate recovered more
quickly in France.

The hospital ships providing this evacuation
service were *Tchad*, *Canada*, *Duguay Trouin*,
Bretagne, *Bien Hoa*, *Vinh Long*, and *Charles Roux*.
Returning from their station off Cape Helles, these

Postcard view of 'French pier,' Mudros.
(Collection of Bernard de Broglio)

ships called at Mudros where they made up their complement of sick passengers before conveying them either to France, or to Algeria or Tunisia, and more rarely to Alexandria. The hospital at Bizerta was reserved more specifically for Algerian troops and Black colonial troops.

During major evacuations, and in the absence of hospital ships, supply vessels were used on their return voyage to France for the repatriation of patients. Patients designated for evacuation were first entered on lists drawn up by the attending medical officers and approved by the senior medical officer of the hospital. Later, repatriation proposals made by the attending medical officers were submitted for review to the hospital's senior medical officer, who conducted a counter-examination, and were then approved by the senior medical officer of the base.

I do not wish to close this chapter without paying reverent tribute to the many soldiers and officers who remain in the soil of Lemnos. Their burial received careful attention, from the French and from our British allies alike, whose dead rest in the same cemetery. A commemorative monument erected in the necropolis will perpetuate their memory for the rare travellers whom chance may bring to this island.[1]

---

1 See appendix.

CHAPTER III

The services

When Evacuation Hospital no. 1 was in full operation—and this was relatively soon after its difficult beginnings—the services were organised as follows:

> General medicine: four fever wards
> Typhoid ward
> Infectious diseases ward
> Surgical department
> Ophthalmology and otorhinolaryngology department
> Bacteriology laboratory
> Radiography laboratory
> Pharmacy.

I.—*General medicine*

This department comprised four fever sections, each containing two to four huts, with an average of 24 beds per hut. Among the conditions treated there, we shall note only those presenting particular features.

Pulmonary diseases.—Pneumonia was frequent, especially among the Black troops (Senegalese and Somalis); in Black patients, susceptibility to the pneumococcus favoured severe forms, with rapid progression towards a fatal outcome. The first damp cold spells of autumn brought the greatest number of these cases.

Pulmonary tuberculosis appeared frequently among the same Black troops, either following pneumonia or bronchopneumonia, or independently of these infections.

Gastrointestinal infections.—Intestinal disorders of every kind accounted for one of the largest contributions to hospital admissions. Apart from infections characterised by specific micro-organisms, which we shall examine among the contagious diseases or in dealing with the typhoid ward, the very large number of febrile gastric disorders that appeared in the expeditionary force must be noted, along with diarrhoea with or without fever. Only these patients were retained in the general medical service. The rise in temperature, when it accompanied these diarrhoeas, was slight and of short duration. The stools were frequently pasty, foetid, and very often blood-streaked. These episodes resolved more or less rapidly with a lacto-vegetarian diet and appropriate treatment. They never gave rise to complications, but one cannot help noting the relationship that appears to have existed between these conditions and cases of jaundice.

Jaundice.—Jaundice began to appear in September. The number of cases increased rapidly, and during

September and October our hospital received 756 cases. The onset was fairly abrupt, and within twenty-four hours the icteric discolouration of the integuments was complete. It was generally not deeply coloured, except in a few cases; its duration was fairly prolonged. Complications were rare, and severe jaundice was encountered only exceptionally. In the patient's history, signs of intestinal infection—so frequent in this army—were almost invariably found, having appeared a longer or shorter time before the jaundice. Many of these patients still had bloody diarrhoea and foetid stools. The condition was accompanied by nervous depression and muscular asthenia, perhaps more marked than those ordinarily observed in jaundice, and of longer duration.

In Black troops, detection of jaundice required careful examination, as it could easily pass unnoticed. It sufficed to test systematically for bile pigments in the urine and to check for icteric discolouration of the sublingual region. The sclerae in these patients normally present a yellowish or brownish colouration that is little altered at the onset of jaundice. Black troops were affected in roughly the same proportion as White troops. They displayed a profound asthenia, though this symptom is habitual among them in even the slightest illnesses. They also complained of pain in the lower limbs, which persisted for a long time.

Dengue.—(Three-day fever, papatasi fever). This

condition proved fairly frequent, especially during the summer months. It is caused by the bite of a small winged insect, the *Phlebotomus*, and also by that of *Culex* mosquitoes. Both insects are found in abundance in these regions. Dengue is well known in Turkey, where frequent epidemics occur in summer.

The clinical picture of this disease is almost always highly characteristic, and we can do no better than borrow from Professor Brumpt's *Précis de parasitologie* the description given by that authority; the symptomatology observed at Mudros corresponded exactly to it.

The disease begins suddenly with articular, muscular, and lumbar pain, cephalalgia, and fever that may reach 40 degrees from the first day. The pain increases in intensity with the slightest movement. The face, mucous membranes, and skin are congested, and a sort of initial rash is observed. Digestive disturbances (anorexia, vomiting, diarrhoea) are frequently noted. Psychic depression is generally very marked.

Around the third day, the patient feels better and, following profuse sweating, appears cured; but this state of well-being, which allows the patient to resume his occupations, lasts only two, three, or four days. It is followed, after this interval, by a relapse characterised by mild fever lasting a few hours and by the appearance of a polymorphous exanthematous eruption persisting for several days.

The illness ends with desquamation of the skin, often accompanied by violent pruritus.

Convalescence from this disease is relatively prolonged, with pain persisting for some time.

Frostbite of the feet.—This condition, which few expected to encounter in this region, followed a spell of cold that struck suddenly in the early days of December 1915. The temperature was exceptional that year: the thermometer fell to 8 degrees below zero, and the old fishermen of Mudros said that within living memory no one had seen such cold.

As is typically observed in such cases, this period of intense cold had to be followed by days of damp weather before frostbite of the feet occurred.

The Black troops were the most severely affected, and while the majority of these frostbite cases had no serious consequences, a number nevertheless progressed to gangrene, necessitating several amputations.

Filariae.—A few cases of dracunculiasis were observed at Evacuation Hospital no. 1. In truth, carriers of the Guinea worm were not hospitalised for this condition. The Senegalese who were affected had, for the most part, already encountered the parasite before, and the extraction they performed was one they had learnt in their own country. Nor did they ask medical science to do better than they could themselves, which would in any case have been difficult. The indigenous method of treatment consists of drawing the worm out through the small open wound it has produced and winding it around a piece of wood, such as a matchstick. Each day, morning and evening, the carrier of the parasite gives the piece of wood several turns. By this somewhat slow but fairly reliable means, he succeeds in extracting three to four centimetres of worm per day. As the filaria measures 50 to 90 centimetres, a mechanical treatment of about a fortnight is sufficient to extract it in its entirety.

Malaria.—Although malaria was observed in our soldiers, it was not on the scale that would become so striking during the Macedonian expedition. There were a few cases at Evacuation Hospital no. 1, but these almost always involved malarial patients who had contracted their infection in Africa or in the colonies; troops from those countries were present in fairly considerable numbers within the expeditionary force.

The rarity of malaria is explained by the corresponding scarcity of anopheles mosquitoes both on the island of Lemnos and at Gallipoli.

As a precautionary measure, however, the medical service prescribed the administration of prophylactic quinine.

II.—*Typhoid service*

Despite the regular administration of anti-typhoid

vaccinations throughout the expeditionary force, these could not prevent the occurrence of a few cases of typhoid fever, for which an isolation ward was established at Evacuation Hospital no. 1.

When the bacteriology laboratory was set up (in the second half of August), it became possible to identify a number of typhoid and paratyphoid fevers among the frequent intestinal infections treated at the hospital. These cases came for the most part from troops operating on the Gallipoli Peninsula; the base formations at Mudros, better protected by a stricter hygiene that was easier to enforce, paid a lesser toll to these infections.

Of the total number of examinations carried out by the laboratory, which served both hospitals and the base simultaneously, we do not have a breakdown of cases relating specifically to Evacuation Hospital no. 1. The examinations comprised 554 blood cultures; 352 yielded a negative result, and 202 were positive. Among the latter there were:

Eberth's bacillus – 56 cases
para-A – 2
para-B – 116
unidentified – 28.

If one considers the distribution of these various infectious agents over time, Eberth's bacillus was observed with greatest frequency during the month of August, whereas from September onwards there was a predominance of paratyphoid B.

The typhoid ward had been equipped from the outset with tarred canvas baths supplied by the naval squadron. These makeshift baths served until those provided by the medical service arrived later, so that balneation could always be applied in the ward.

The cerebral form was observed on a number of occasions, especially during the period of intense heat. It manifested as severe cerebral excitation and ambulatory delirium. The proximity of the sea seemed to draw these patients towards the water, and more than once staff could be seen chasing through the hospital after a patient who, eluding the otherwise vigilant supervision of the staff, rushed towards the beach. No accidents resulted from these incidents.

Together with the dysenteries, typhoid fever may be counted among the diseases that caused the greatest losses at Evacuation Hospital no. 1.

III.—*Infectious diseases ward*

This ward received all patients suffering from contagious diseases, with the exception of typhoid cases.

Situated at the far end of the hospital, it comprised four huts, a masonry building reserved for the most seriously ill patients—to whom it offered greater comfort—and *tentes tortoises* and *marabout* tents used to isolate the most contagious cases. Its capacity was approximately 150 patients.

The contagious diseases treated in this ward, in addition to the exceedingly numerous cases of dysentery, were as follows:

diphtheria – 6 cases
cerebrospinal meningitis – 1
measles – 3
scarlet fever – 1
tetanus – 2
mumps – 5.

The isolation of these patients was carried out in *marabout* tents set at a distance from the huts and spaced apart from one another. This isolation proved effective, for neither eruptive fevers, nor diphtheria, nor mumps gave rise to any epidemic focus within the hospital. Disinfection of the *marabout* tents and bedding was easy to carry out. It was always applied methodically; but above all, ventilation and exposure to sunlight certainly played an important part on an island ceaselessly swept by winds from the open sea and under constant sunshine.

Dysentery accounted for a large proportion of morbidity at the Dardanelles. One cause of its spread lay in the extreme abundance of flies. Protection of premises by means of wire mesh on windows and vestibule screens on doors was adopted whenever possible. Likewise, disinfection of residues of every kind was carefully carried out. During periods of intense heat, however, patients were housed chiefly under tents, which at that time were the only shelters available: flies could accumulate in them freely. Methods of destruction such as flypapers, formaldehyde-treated milk, cresol washes, and so forth wreaked considerable slaughter, but these losses counted for little: the insects bred so prodigiously that the air itself seemed always thick with them. Contagion within the hospital itself was only a very minor addition to the considerable number of dysentery cases arriving from Sedd-el-Bahr. On the peninsula, flies were at least as abundant, and protecting the troops against the contagion they carried was a material impossibility.

From a clinical standpoint, the dysenteries observed at Evacuation Hospital no. 1 were remarkable only for their low severity. This was probably because they were chiefly cases of bacillary dysentery. This conclusion, based on clinical findings, was confirmed by the laboratory. There was in fact a remarkable scarcity of complications in the form of liver abscess; the few hepatic abscesses requiring surgical intervention occurred in colonial troops, and everything suggests that these men had brought with them amoebae acquired previously, rather than having contracted them at the Dardanelles. The unquestionable efficacy of treatment with antidysenteric serum—widely administered from the earliest observed cases and continued regularly thereafter—strongly suggested the presence of dysentery bacilli. Finally, the laboratory was able to identify amoebae in only two cases; it is fair to note, however, that this laboratory did not begin operating until August, that is to say at a time when severe dysentery cases had already diminished in number; but even then it was able to confirm bacillary dysentery in many cases.

Neither plague, nor cholera, nor typhus was ever observed at Evacuation Hospital no. 1.

IV.—Surgical service

This service comprised two divisions:

The first, reserved for the seriously wounded.

The second division, for all other wounded, also hospitalised patients suffering from conditions of the eyes, nose, throat, and ears. These patients, however, received care from a specialist physician. The ophthalmology and otorhinolaryngology service will be described in a later section.

These two surgical divisions had at their disposal various dressing and examination rooms, and shared the same surgical block.

The surgical block was laid out as follows. A dressing room, with no connection to the rooms in which operations were performed, opened onto the hospital courtyard through a small room serving as a cloakroom and waiting area.

Adjacent to this dressing room, with its own separate entrance, was the preparation room in which the wounded were prepared and, where necessary, anaesthetised before surgery. A small compartment within this room housed an Adnet autoclave,[1] two Primus stoves, and the receptacles used for sterilising instruments.

From this room one entered the operating theatre. Well lit, receiving daylight from three sides, entirely finished in white Ripolin enamel,[2] its floor covered with linoleum, equipped with a folding operating table of lacquered metal and the indispensable array of receptacles for solutions and dressings, it did credit to the service and allowed a great deal of useful surgical work to be carried out.

More than 300 operations of all kinds were performed between 28 April 1915 and 1 February 1916. The wounds most frequently observed at Evacuation Hospital no. 1 were perforating bullet wounds. In general, all wounds were infected on arrival; they were consequently often followed by suppuration, and in a fairly high proportion of cases necessitated secondary procedures.

Bone injuries were also numerous: crushing of bone by shell fragments and shattering of bone tissue following gunshot wounds.

1 Adnet autoclave: a pressurised steam steriliser used for the sterilisation of surgical instruments and dressings—the standard method in French military surgical units by this period.

2 Ripolin: a proprietary white enamel paint widely used in French hospitals and surgical facilities for its hard, washable surface.

Wound infection was in general attributable to the presence of foreign bodies—fragments of clothing, projectiles retained in the tissues—or of bone fragments, splinters, and sequestra. This infection, spreading to the lymphatic territories, was frequently followed by abscesses, phlegmons, and extensive tissue separation. Gas gangrene manifested in several cases involving irregular wounds, notably in crush injuries of bone with deep tissue destruction.

Tetanus was very rarely observed, the wounded having always received anti-tetanus serum immediately after being wounded.

Vascular injuries (aneurysms, and arterial or venous section) were rare, notwithstanding the number of perforating bullet wounds affecting the limbs.

Abdominal surgery was not often performed at Evacuation Hospital no. 1.

The distance of this hospital from the theatre of military operations precluded the direct admission of casualties with severe injuries to the intestine, liver, stomach, and the like. Such cases were received and operated upon at an earlier stage aboard the hospital ships stationed off Cape Helles. Nevertheless, certain interventions became necessary upon the liver, for large amoebic abscesses. Similarly, a number of laparotomies were performed as a matter of urgency for intraperitoneal purulent collections resulting from intestinal perforations of typhoid origin.

Cranial wounds were relatively few, particularly during the first four months of the hospital's operation. They required in some cases urgent trepanation, or more often debridement of the fracture site with removal of bone splinters.

In sum, the surgical block, equipped with the essential surgical instruments and the sterilisation apparatus necessary for asepsis, permitted the performance of all requisite operations.

The contribution of the radiography laboratory proved valuable; unfortunately it was not established within the hospital until the second phase of operations, towards the end of the summer.

V.—Ophthalmology and otorhinolaryngology service

For several months, these patients were hospitalised in a division of the surgical service.

The ophthalmology service was set up at the beginning of June. A *tente tortoise* served as its first premises, doubling as both consulting room and operating theatre, open to the wind on all sides. A few packing cases and planks were its first furnishings. Drugs and instruments were rudimentary. The equipment provided to the hospital—intended only as a place of transit—comprised a few drugs for routine ophthalmological use and little instrumentation. These were supplemented by what the senior medical officer in charge of the service had brought with him.

In the absence of a darkroom—difficult to set up

beneath the canvas of a tent—ophthalmoscopic examinations could only be performed at night, by the light of a small paraffin lamp.

Towards the middle of August a hut was built for this service, in which a consulting room and a more comfortable operating theatre could be set up. A second hut was soon added, used to hospitalise the wounded or the sick undergoing treatment. Little by little, through officers returning from France or Egypt, the stock of small equipment grew and was supplemented: instruments, spectacles, dressings.

The ophthalmology and otorhinolaryngology service treated a total of 341 patients with disorders of the nose, throat and ears, and 833 patients with ocular disorders. The latter were distributed as follows:

various conjunctivitis – 305
granular conjunctivitis – 110
myopia with fundus lesions – 56
corneal ulcers – 123
various war injuries – 175
miscellaneous conditions – 64.

The ophthalmological operations performed were:

25 scrapings
15 trichiasis operations
37 iridectomies
59 extractions of foreign body
15 cataract operations
6 pterygium excisions.

VI.—*Bacteriology laboratory*

Established to carry out the bacteriological investigations required by the medical formations— Evacuation Hospitals nos. 1 and 2, and Convalescent Hospital no. 2—and by the hygiene service of Mudros base, the bacteriology laboratory was set up on 20 August 1915 at Evacuation Hospital no. 1.

The equipment was supplied by the laboratory of the expeditionary corps at Sedd-el-Bahr, which handed over to it such culture media and apparatus as it could spare. However generous the consignment from Cape Helles may have been, the statistics of analyses performed at the Mudros laboratory will show that the necessarily rudimentary equipment at its disposal was somewhat inadequate relative to the volume of examination requests directed to it.

The staff comprised one medical officer and two orderlies.

Statistics of analyses.—Blood culture for the diagnosis of typhoid states, stool examinations for the diagnosis of dysentery, and analyses of blood, sputum, false membranes and cerebrospinal fluid constituted the principal categories of investigations carried out by the laboratory.

Blood cultures.—The laboratory performed 554 blood cultures on patients arriving either from Sedd-el-Bahr or from the base, most commonly with a diagnosis of febrile myalgia. 202 blood cultures were positive and 352 negative.

Identification of the bacilli isolated in the 202 positive cases revealed the presence of the Eberth bacillus on 56 occasions, paratyphoid B on 116 occasions, and paratyphoid A on only two occasions.

Twenty-eight strains of bacilli isolated from patients' blood could not be precisely identified, owing to the laboratory's shortage of culture media. Most of these were, in any case, bacilli belonging to the Eberth–paratyphoid B group.

In four cases, either cocci or a bacillus were observed that appeared analogous to those described by Besredka and Sacquépée.[1]

A remarkable finding was that the Eberth bacillus reached its maximum frequency in August and September and became exceptional from October onwards. The inverse proportion was observed for paratyphoid B. The rarity of paratyphoid A is also worth remarking: only two cases, observed on the same day. Their identification was completed by specific agglutination. At the same time, the naval laboratory at Mudros likewise identified three cases of paratyphoid A.

---

1 Alexandre Besredka and Émile Sacquépée were bacteriologists at the Institut Pasteur, Paris, who collaborated on research into the bacterial agents of dysentery and related infectious diseases.

Search for haematozoa.—The laboratory investigated haematozoa under two different conditions:

a) In the blood of 40 patients at Evacuation Hospital no. 1 who had been identified as possible malaria cases, the majority belonging to the home army. In only seven cases was the result positive. Two of these soldiers had never previously left France; the other five had served in or were resident in Algeria or Morocco. It should be added that in the 33 negative cases, blood culture carried out concurrently revealed the presence of a typhoid state.

b) In the blood of 40 febrile patients with negative blood culture. The search for haematozoa in these cases was invariably negative.

A concurrent search for anopheline larvae yielded only a few specimens, and these confined to certain parts of the island (Parthénontos, Prétoria).[1] Elsewhere the larvae encountered were those of *Culex*.

It should also be noted that the same investigations carried out by the laboratory at Sedd-el-Bahr showed that malaria was very little in evidence at the Dardanelles.

---

1 Parthénontos is probably near modern-day Parthenomytos. Prétoria (sometimes Prétaria) has not been identified; it may be a name given to a military encampment by Allied forces, or a rendering of a local place name not currently traceable.

Examination of stools.—The search for amoebae yielded a positive result in only two cases.

As for the bacteriological examination of stools, this could not be undertaken owing to lack of equipment.

However, in November the laboratory had to analyse the stools of Turkish prisoners suffering from very severe diarrhoeal disorders. The search for the cholera vibrio was negative. By contrast, the presence of dysenteric bacilli made it possible to direct treatment along favourable lines. This confirmed the clinical experience previously acquired, and justified treatment with antidysenteric serum, applied with success to the numerous dysentery patients who had filled the hospital from its foundation and who had been unable to undergo analysis for want of a laboratory at that time.

Bacteriological analysis of water supplies.— Seventeen complete water analyses were carried out by the following methods:

bacterial count
colimetry
search for putrefactive organisms.

Practical results as applied to water consumption:

a) springs at Lichna and Varos: potable (no coliforms)

b) water distilled by the *Shamrock* and the *Chasseloup-Laubat*:[1] potable, no coliforms, owing to chlorination

c) all other well or spring water in the vicinity of our units was non-potable.

Miscellaneous investigations:

Diphtheria: 3 positive cases.

Cerebrospinal fluid:

1 case of meningococcus
1 case of tuberculosis
1 case of streptococcus.

Sputum: 150 examinations, with 3 positive cases.

1 The *Shamrock* was an Annamite-class transport (1878), and the *Chasseloup-Laubat* a decommissioned Friant-class cruiser (1893), both recommissioned in 1915 and anchored at Mudros as part of the French Navy's *Groupe des navires affectés au service de l'eau*. The *Chasseloup-Laubat*'s coal-fired boilers were converted into a floating distillery *(usine distillatoire)*; the *Shamrock* served in the water group and doubled as a hospital transport with an on-board bacteriology laboratory.

VII.—*Radiography laboratory*

Established relatively late in the hospital's existence, the laboratory was nonetheless well equipped and served to carry out a considerable number of radiographic and fluoroscopic examinations.

Beyond its proper functions, and this proved a valuable resource during the winter months, the laboratory's generating set supplied electric light to the entire hospital.

VIII.—*Chaplaincy*

Three ministers of religion held officer rank on the hospital's establishment: a Catholic chaplain, a Protestant pastor, and a rabbi.

Given the larger number of Catholic soldiers, a wooden chapel was built. The other services were held in turn in the *marabout* reserved for this purpose.

IX.—*Pharmacy*

On its arrival at Mudros, Evacuation Hospital no. 1 had at its disposal, in the way of medicines, only:

> two panniers no. 1 [1]
> two panniers no. 12 [2]
> two cases of sera (anti-tetanus, anti-dysenteric, etc.)
> two cases of disinfectant materials.

---

1 Surgical instruments and bandages.

2 A portable field pharmacy: morphine, quinine, chloroform, iodine, and other medications with dispensing equipment.

These resources were slender indeed in view of the number of sick and wounded the hospital was about to receive. At that stage, however, it was still intended to serve merely as a staging post between the forward medical formations and the base hospitals proper, and had only 200 stretchers at its disposal.

The medical officer in command, foreseeing the hospital's future role, nevertheless dispatched an urgent request for medicines to the Ministry. Mudros as yet had no pharmaceutical reserve. As a precaution—fully justified by events—steps were taken to obtain local supplies while awaiting the outcome of this request, and above all to save the time that would be lost given the distance from France and the difficulties of transport. There were two pharmacists at Kastro, the principal town of Lemnos. A quantity of medicines was obtained from them, and through their offices the town's merchants supplied 95 per cent alcohol. Traders brought further medicines from Salonika, including several kilograms of iodine. On these resources the hospital was able to supply medicines to its patients for nearly three months. In July it received the order dispatched to the Ministry in May. From that point the era of pharmaceutical scarcity came to an end, and in August something approaching abundance arrived when the expeditionary corps' pharmaceutical reserve was established at the Mudros base.

The pharmacy's first premises were modest; there was hardly need for much space! A *marabout* tent and a cellar sufficed. The cellar, cut into the hillside, housed inflammable materials—alcohol, ether, chloroform and the like—together with products susceptible to heat: sera, vaccines, sticking plaster, and so on.

When the base was later able to supply timber, and Turkish prisoners were made available as labour, a large hut twelve metres long by six metres wide was built and fitted out. The pharmacy, now comfortably housed and well stocked, was able to maintain a regular service. Through a simplification of prescriptions and formulae—the list being sent to the pharmacy immediately after the ward round—medicines were dispensed at one o'clock in the afternoon. Distribution had known other difficulties earlier, when—in the absence of phials—medicines had to be issued in condensed milk tins. These were later replaced by mineral water bottles, and finally by regulation phials.

The dispensary for infusions and tisanes was housed in the kitchens and distributions were made morning and evening.

Water supply.—From the moment of landing, one of the hospital's first concerns was to secure a supply of drinking water. A nearby spring was tapped and its water—approximately 1,500 litres a day—piped to a storage barrel by means of a wooden conduit.

For a time this spring met the needs of the staff, but by the end of April, with the arrival of the first patients, it proved insufficient. No other springs being available in the immediate or more distant vicinity, wells were sunk in the Mudros plain by Turkish prisoners. For want of building materials and skilled labour, however, these were no more than simple excavations from which the water was drawn off during the day. The Mudros plain lies at sea level, and these holes being close to the shore, they filled overnight with brackish water. Furthermore, the walls being supported only by stone or timber, the water was already cloudy at the first drawing and muddy by evening.

Water was transported to the hospital by araba, the cart carrying a 600-litre barrel. A second araba, provided somewhat later, brought the daily supply to approximately 4,000 litres. This meagre quantity, given the large patient population and considerable staff, maintained the hospital's water service for several months. The water was far from clear and unpleasant to the taste despite treatment.

Spring water, which was agreeable to drink, was reserved for the preparation of milk and infusions. Well water was used in the kitchens, for drinking, washing, and cleaning. The equitable and careful distribution of so small a quantity of potable water was a genuine problem, particularly during the great heat. In the wards, only bedridden patients had unrestricted access to drinking water; the others—the walking wounded and nursing staff—were required to come to the pharmacy water point and were entitled to no more than a quarter-litre at a time. The filling of water bottles was forbidden. This point was supervised by an orderly, and the pharmacist himself oversaw the careful use of the precious liquid.

For the baths, only seawater was used.

Towards November, the base installed a pipeline to bring to the hospital the sterilised water produced by distillation aboard the *Shamrock* at Mudros. Thanks to this somewhat belated addition, the hospital was at last adequately supplied with drinking water.

Water sterilisation.—The water from the first spring tapped near the hospital, with a daily yield of approximately 1,500 litres, was sterilised by the Lambert process. The installation consisted of a bank of six Garet filters set up in the open air.[1]

Well water was initially sterilised by the same process before chlorination was introduced. The

1 The Lambert process was a French proprietary water purification method, most probably based on ozonation, which leaves no chemical residue in treated water—a significant practical advantage over contemporary chlorine and iodine treatments. The Garet filters were likely ceramic pre-filters of the candle type, used to clarify water before sterilisation; such two-stage arrangements, filtering then sterilising, were standard practice in French military and civilian installations of the period.

installation comprised a bank of four barrels of approximately 600 litres each, in which the water was treated, and a bank of six Garet filters for filtration before consumption. Both banks were housed in a timber shed.

Shortly afterwards the Lambert process was abandoned in favour of chlorination. For this purpose, cement cisterns were installed near the kitchens. These cisterns had sloping floors to allow settling and cleaning. Each had two taps at the lower end, one above the other: the higher tap drew off water for consumption, while the lower could drain the cistern completely of its sediment and washing water. To ensure thorough settling, the process was repeated twice in succession on the same water. A first cistern received the raw water, to which the necessary quantity of Javel extract was added;[1] chlorination was then allowed to act for a minimum of six hours. The water was then drawn off through the upper tap for a first settling. A lower cistern received this already sterilised water, where it was left to settle for a further few hours before being passed for consumption.

---

1   Javel extract: a solution of sodium hypochlorite, used as a disinfectant and water purification agent. Named after the former Parisian village of Javel where it was first produced in the late eighteenth century; the equivalent of household bleach.

Disinfection service.—In the early period of the hospital's operation, soiled linen was steeped in formalin solution in wash-boilers, then washed with soap and dried in the sun.

For the destruction of parasites, blankets hung on lines were sprayed with formalin and then subjected to a hot iron.

Later, the disinfection of linen, blankets, and mattresses was carried out in a large, wheeled steam disinfector.

Two brick incinerators, one at each end of the hospital at a distance of approximately one hundred metres from the huts, disposed of all hospital waste.

Ice.—At the same time as the request for medicines was dispatched to the Ministry on the hospital's establishment, a request was also made for an ice-making machine. This arrived in July but proved to be a small apparatus; the hospital at that point had 1,200 patients, including a large number of typhoid cases. The machine operated on sulphuric acid, driven by a hand pump, and could at best produce a single chilled carafe after approximately an hour's operation, provided the temperature was sufficiently low—a condition difficult to meet in a climate where the thermometer held between 35 and 40 degrees. Despite the resigned efforts of the Turkish prisoners set to this work, the machine never yielded so much as a single piece of ice.

Fortunately the hospital held funds from donations, and from these a sum was drawn to procure from France a more substantial machine. This one, costing 2,500 francs, operated on carbonic acid and was driven by a 6 h.p. petrol engine, with a daily output of 100 to 150 kilograms of ice. Though it had no official status, it supplied ice continuously to Evacuation Hospital no. 1 (1,500 beds), Field Hospital no. 2 at Lichna (1,000 beds), and Evacuation Hospital no. 2 at Lichna (1,000 beds) from August 1915 to February 1916. At the price at which the Commissariat supplied petrol for the engine, the cost of production was ten *centimes* per kilogram. The sole drawback of the machine was the quantity of water required to cool the engine and the carbonic acid compression pump—three cubic metres per day, at a time when the hospital's total daily supply was only 4,000 to 5,000 litres. A water recovery system was therefore installed to reduce losses to a minimum.

The pharmacy staff comprised:

Two pharmacists: one responsible for the pharmacy proper, the other for analyses and general hygiene.

1 corporal orderly and 1 orderly, qualified pharmacists or students, covering night duty

1 orderly for infusions and chlorination

1 orderly-mechanic for the ice machine and water supply

1 orderly for disinfection.

X.—*Diet and kitchens*

The feeding of patients at Evacuation Hospital no. 1 was maintained in a satisfactory manner through three channels:

1. the military commissariat at Mudros

2. the medical service reserve stores

3. local resources. These were varied in origin. The island of Lemnos offered some produce, which was duly exploited, though in modest quantities. Supply vessels were called upon and contributed greatly to meeting requirements. When a regular shipping service linked Mudros with Salonika—where the Army of Macedonia had landed in September 1915—that city became a further significant source of supply. It was still a period when purchases could be made easily and cheaply in a large centre as yet unexploited, though Salonika was soon to become one of the most expensive places in the theatre.

The dietary régime at Evacuation Hospital no. 1 differed entirely from that prescribed by military regulations for evacuation hospitals proper. The hospital retained large numbers of patients requiring varied diets, and the standard ration scale was unsuited to cases of every category. In their interests, the medical officer in command did not hesitate to substitute the dietary regulations of the home medical service for those of the field medical service, and at the same time made the keeping of ward visit books obligatory.

The dietary categories were as follows: full diet; light diet; special diet; restricted diet (milk diet or absolute fast).

The full diet requires no particular comment. The light diet could be composed of meat and vegetables—in which case it was termed the *mixed light diet*—or of two vegetable courses, in which case it was known as the *vegetable light diet*. The special diet allowed two items from the tariff; these, more varied than those of the light diet, might include eggs, vegetables, puréed vegetables, or desserts.

All these diets were governed by menus drawn up in advance for a week at a time. Every possible variety was introduced into the composition of these menus, limited only by local resources and economic conditions.

Diet was of particular importance in a hospital environment where disorders of the digestive tract were prevalent. It was a constant preoccupation of the medical staff, and this aspect of treatment proved a valuable complement to the therapeutic management of such conditions.

———

Mudros scenes.

(Elisabeth Jardin Collection)

CONCLUSIONS

From the study of the operation of Evacuation Hospital no. 1 at Mudros, the following conclusions may be drawn:

1. The most frequent and most serious conditions treated in this hospital were those of the digestive tract: a certain number of typhoid cases (Eberth and para-B, and exceptionally para-A), and above all a very large number of dysenteric conditions. In the latter, laboratory examination established the rarity of amoebic dysentery, but in contrast a very high frequency of dysentery bacilli. The extensive use of anti-dysentery serum from the outset of the Dardanelles expedition, and the good results obtained from it, had already permitted this clinical conclusion to be drawn several months before laboratory confirmation became available.

2. Jaundice made its appearance in the expeditionary corps with the first cold damp weather of September. It appeared very rapidly, as a veritable epidemic, and the number of cases was very high. The course of the illness was straightforward and uncomplicated. Black troops appear to have been affected in roughly equal proportion to White troops. It is to be regretted that during the course of these jaundice cases, the laboratory—barely

established and very summarily equipped—was unable to undertake investigations to detect ictero-haemorrhagic spirochaetosis.

3. Pulmonary conditions were frequently observed, and in large numbers among African troops (Senegalese and Somalis). In the latter there was a predominance of severe pneumonia and pulmonary tuberculosis.

4. Numerous cases of frostbite of the feet were observed, particularly among African troops. A brief cold spell followed by a period of damp proved sufficient to cause these injuries. Notable was the exceptional fall in temperature in the Aegean during the winter of 1915: the thermometer dropped to 10 degrees in the first days of December.

5. Dengue fever (three-day fever, pappataci fever) was frequently observed during the summer months. This condition, very common in Turkey, was not serious in character.

6. Malaria, by contrast, was encountered only exceptionally. The few cases noted were in soldiers who had served in the colonies or in Africa, colonial and African troops forming part of the expeditionary corps establishment.

7. The presence of Senegalese troops afforded the opportunity to observe the extraction of the Guinea worm by the native method, carried out by those harbouring the parasite.

8. Contagious diseases were rarely encountered, and Evacuation Hospital no. 1 had no cases of plague or cholera to treat.

9. Evacuation Hospital no. 1 at Mudros, despite the designation it retained for administrative purposes, functioned in practice as a fully constituted *hôpital temporaire*, with up to 1,500 sick and wounded under treatment.

* * *

APPENDIX

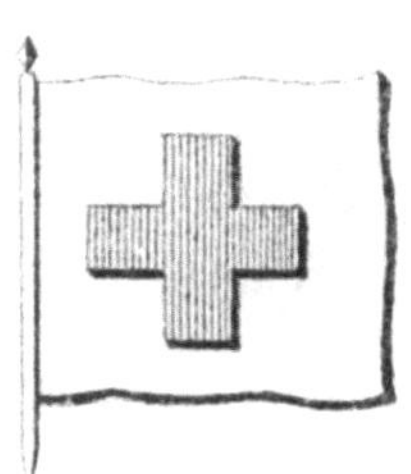

FRENCH HOSPITALS AT GALLIPOLI 1915/16

This survey was first published in 'The Gallipolian,' the journal of the Gallipoli Association (number 165, 2024). The translator has updated the text, which attempts to document the field hospitals of the French expeditionary force.

France's military medical services

To start, some background on France's army medical corps *(Service de santé des armées)* may be helpful.

The mobilisation of 1914 was carried out according to a 1910 plan that distinguished between front *(zone des armées)* and rear areas *(zone de l'intérieur)*, with a chain of evacuation familiar to students of British and Dominion medical services: regimental stretcher-bearers carried wounded to a regimental aid post, from where divisional stretcher-bearers used horse-drawn transport to reach a field ambulance. Motor transport then moved the casualties to an evacuation hospital (what Imperial forces called a casualty clearing station) at or near a railway line, for evacuation to the rear and further treatment in a military, auxiliary or civilian hospital.

Medical officers were drawn from both serving military personnel and civilian reservists. France had universal military conscription as a condition of citizenship and reservists made up about 90% of the medical personnel mobilised for the Great War.

Medical students who had yet to complete their training held the rank of *médecin auxiliaire* or *médecin sous-aide major*, equivalent to a non-commissioned officer (warrant officer or chief warrant officer). Having qualified, the doctor was appointed *médecin aide major de 2ème classe*, a junior officer rank equivalent to a second lieutenant.

Before the First World War, army doctors rose through the ranks much faster than reservists who rarely obtained a rank higher than *médecin major de 2ème classe* (captain). Army doctors also had seniority over a reservist of the same rank. The want of medical officers in wartime soon eliminated these distinctions.

As for trained nurses, the French army had less than 9,000 in August 1914. At this time, all military nurses were male, although some military hospitals employed female lay nurses. About 100,000 male nurses and orderlies were mobilised at the onset of war to staff stretcher-bearer groups, field ambulances and evacuation hospitals, but almost half had no paramedic training. More than 10% were clergymen, forced into exile by anti-congregational laws, who returned to France when their classes were called.

A role for trained female nurses in wartime had long been anticipated, but as volunteers. France had no less than three Red Cross societies. They competed for money, patronage and prestige, although their vocation cannot be questioned. The most senior of the Red Cross societies was founded in 1864. The *Société de secours aux blessés militaires* (SSBM, Society to Aid Military Wounded) was conservative and Catholic. The *Association des dames françaises* (ADF), founded 1879, was republican and anticlerical, and the *Union des femmes françaises* (UFF), founded 1881, was feminist and Protestant. In 1914, some 30,000 nurses were mobilised from the Red Cross, with 72,000 effective in 1918.

Fundraising stamps issued by French Red Cross societies.
Left: 'The Dardanelles', *Société de secours aux blessés militaires. Right:* CEO commander General Gouraud on a stamp issued by the *Association des dames françaises.*

The need for a professional cohort of female military nurses became obvious early in the war. In 1916, a temporary paid position was created for female nurses in the French army, followed by a formal military nursing role in 1918.

The French Navy's medical service, the *Service de santé de la Marine*, operated as a distinct entity, managing an infrastructure that spanned shore-based hospitals, hospital ships, and field units attached to the naval infantry. Specialising in maritime evacuation, the service commenced the war with fifty doctors stationed in the Mediterranean.

During the Great War, the medical complement of a French hospital ship typically consisted of a senior medical officer, three or four medical officers (including a surgeon), three or four junior physicians, and one or two pharmacists. Nursing duties were performed by 16 to 20 male nurses and a variable detachment of female Red Cross staff, while sailors with basic medical training served as orderlies. These vessels were equipped with an operating theatre, a pathology laboratory, and an X-ray department.

The French expeditionary force at Gallipoli

The French contingent to join General Sir Ian Hamilton's Mediterranean Expeditionary Force was formed by order of the French Minister of War on 22 February 1915.

The *Corps expéditionnaire d'Orient* (CEO), initially comprising one division, began embarking for the island of Lemnos on 4 March. Due to the lack of port facilities, most of the division was shifted to Alexandria in late March, although its hospital remained on Lemnos.

On 25 and 26 April, the French made a successful diversionary landing on the Asia Minor shore at Kum Kale, before transferring their forces to Helles, where they took the right of the line. On 29 April 1915, a second infantry division was raised as a reinforcement. It began landing at Helles on 6 May.

The *Corps expéditionnaire d'Orient* (CEO) became the *Corps expéditionnaire des Dardanelles* (CED) on 4 October 1915 when the 2nd Division left the Gallipoli Peninsula for Salonika and operations in Macedonia. The 1st Division remained at Helles until its last elements were evacuated on 9 January 1916. The CED was dissolved, and the 1st Division was regrouped as the 17th Colonial Infantry Division at Mytilene (Lesbos). On 15 February 1916, this division was landed at Salonika.

CEO medical services

Each French division at Gallipoli was equipped with an evacuation hospital, a field hospital, two field ambulances and a group of divisional stretcher-bearers. (Regiments also carried a stretcher-bearer group, whose responsibility was to clear the frontline zone.)

The CEO had a chief medical officer as part of its headquarters staff. So too did each division, as well as Mudros base, the CEO's principal Lines of Communication base.

The French navy had anticipated the need for hospital ships to accompany the CEO but was forced to requisition more ships when the Allies were restricted to shallow positions at Helles and Anzac, and it became clear that base hospitals could not be landed on the peninsula.

The navy also established an infectious diseases hospital on shore at Lemnos.

The history of French medical services at Gallipoli remains fragmented, even within French historiography. However, recent scholarship—notably the valuable directory compiled by François Olier and Jean-Luc Quenec'hdu—is beginning to address this deficit. Furthermore, a wealth of material is emerging from the digitised collections of the armed forces, the National Library of France and provincial archives. While further research is required, this provisional list of French hospitals at Gallipoli offers a starting point.

Hospitals on the Gallipoli Peninsula

Field Hospital (Hôpital de campagne) no. 1

This hospital was established in the northern part of Sedd-el-Bahr castle (*Château d'Europe*), with 88 beds. From 29 April until 10/11 May, when

Field Ambulance no. 1 was landed, this was the only French hospital on the Gallipoli Peninsula. (Desperate fighting took place in early May when Turkish counter-attacks pushed French forces back to the heights above Morto Bay.)

Sedd-el-Bahr was exposed to Ottoman artillery throughout the campaign. A senate commission that inspected the hospital described it as 'working under fire.' Only seriously wounded men were admitted; all others, including sick, were evacuated by hospital ship. Lightly wounded were transferred by a regular ferry service to Lemnos.

Although designated a field hospital, the French official history states that it was functioning as an evacuation hospital (casualty clearing station).

Mention here should also be made of the valuable pathology work performed by French medical personnel, who carried out practically all laboratory work for the British until their facilities were landed. The laboratories were established in basements in Sedd-el-Bahr village and in the castle itself. 'They were extremely well equipped,' wrote a British doctor, 'with all possible necessities sent from the Pasteur Institute in Paris.'

Field Hospital no. 1 moved to Tenedos on 21/22 September 1915.

Field Ambulances nos. 1 to 4

Ambulance no. 1, carried aboard *Savoie*, participated in the Kum Kale landing on 25/26 April 1915, then opened at Sedd-el-Bahr castle, in the courtyard of a former Turkish hospital, on 11 May. Capacity was 97, in tents and among the ruins of the castle.

Ambulance no. 2 opened on 13 June in the lower courtyard of Sedd-el-Bahr castle, in tents (190 beds), then hutted barracks (266 beds).

Ambulance no. 3 opened on 22 May in the upper courtyard of Sedd-el-Bahr castle, where there once was an olive oil manufacturing plant, in tents (50 beds). It took the place of Field Hospital no. 1 when that unit moved to Tenedos in September 1915.

Ambulance no. 4 opened on 13 July at the entrance to Sedd-el-Bahr village, in tents (74 beds).

Field Ambulance no. 2 in the lower course of Sedd-el-Bahr castle.
(Sir Ian Hamilton Collection, King's College London)

Hospitals on the island of Lemnos

Lemnos became the principal forward medical base for the French army, as it did for British and Anzac forces. On 24 June 1915, there were 2,400 French wounded and sick in the hospitals ashore and 1,000 on hospital ships in the harbour.

Greek hospital (Hôpital grec)

This small hospital with 20 beds was used by the French expeditionary force between 15 and 24 March 1915, after which patients were transferred to Evacuation Hospital no. 1. The Greek hospital was in Mudros town, beside the post office, near the waterfront.

Evacuation Hospital (Hôpital d'evacuation) no. 1

This hospital embarked with the 1st Division of the CEO, reaching the island of Lemnos on 15 March 1915. It opened on 26 March, with 200 beds, perched on the side of a barren hillside, 1.5 kms (1 mile) south of Mudros, with expansive views over the harbour and town.

Evacuation Hospital no. 1 became the largest hospital on the island. In October 1915, the hospital had 1,500 beds, but total sick and wounded may have reached as many as 1,800. In his 1920 memoir, Doctor Léon Peaudeleu, *médecin aide major de 1ère classe* (lieutenant), gives a total capacity of 1,700. For comparison, the largest British and Dominion hospital was the 3rd Australian General Hospital which opened in August 1915 with 1,040 beds. A similarly-sized British general hospital, No. 27, did not arrive until November.

'At first,' writes Dr Peaudeleu, 'everything was in short supply and it was impossible to get anything on the island. Water was especially hard to come by... sick and wounded were housed in tents on straw bedding ... a violent wind would turn tents inside out like umbrellas. Wood was scarce on the island so huts had to wait until supplies arrived from France.'

Nevertheless, extensive building and other works began soon after the hospital was landed at Mudros. A track from the town to the hospital was improved to become a road, the ground for the hospital was terraced, levelled and drained, and wells were sunk, a reservoir dug and a water supply installed. The construction of an operating theatre began and shelters for kitchens were erected. By September, the hospital comprised 14 large barrack-like huts and 143 tents.

Among the medical staff of the hospital were female nursing sisters. The bulletin of the *Société de secours aux blessés militaires* gives us the names of eight Red Cross nurses. The matron was *Mademoiselle* Yolande Oberkampf. The nursing sisters were *mesdames* Bouchez, de Brinville, Marie Laporte, Jeanne Noblemaire (née Antelme), and Romain-Bougère, and *mademoiselles* Elisabeth Jardin, and Lacaze.

They arrived in two echelons. The first group, including Oberkampf and Jardin, arrived in July 1915 aboard the hospital ship *Duguay Trouin*. They were reinforced in August by a second group, including Jeanne Antelme, who arrived aboard the hospital ship *Charles Roux*.

Near the hospital was a prisoner of war camp. Ottoman soldiers were employed on earthworks and other construction tasks for the hospital. In the summer, Dr Peaudeleu says that the prisoners suffered from dysentery, typhoid and paratyphoid fevers, as well as complications from their wounds.

Evacuation Hospital no. 1 departed Mudros in February 1916 aboard *Tchad*.

'Despite the isolation and destitution in which we found ourselves,' wrote Dr Peaudeleu, 'more than 20,000 sick and wounded had found asylum in this corner of the world, which represented distant France for them.'

The hospital would serve at Salonika as Auxiliary Hospital *(Hôpital temporaire)* no. 7.

Evacuation Hospital no. 1, overlooking Mudros harbour.
The town is seen top-right.

*(Ernest Brooks, Admiralty official photographer,
composite Bernard de Broglio)*

Infirmary (Infirmerie-hôpital)

An annexe of Evacuation Hospital no. 1, this tented hospital was located on the flat coastal fringe about 1.5 kms (1 mile) southwest of Mudros town, not far from the main hospital. Initially designated *infirmerie-ambulance*, it was upgraded to *infirmerie-hôpital* in August 1915. Capacity was 150 to 180 beds. The infirmary admitted lightly wounded men, as well as sick (parasitic and venereal diseases).

Apparently this facility also administered a convalescent camp *(dépôt d'éclopés)*, 5 km (3 miles) from Mudros town, which opened on 14 June 1915. This may be the convalescent camp on the Bay of Pournia, north of Varos, although the distance is greater than 5 km. Establishing works on the Pournia camp were underway in June, and it was receiving patients by 23 September 1915.

Field Hospital (Hôpital de campagne) no. 2

The 2nd Division's field hospital opened on 15 June 1915 near the village of Lychna, on the harbour's shore about 4 kilometres (2.5 miles) north of Mudros on the road to Kastro (modern-day Myrina). By September 1915, it consisted of seven barrack huts and 93 tents. Two huts were built to accommodate Red Cross nurses but it is unclear if any joined the hospital before it closed. (In August, the medical HQ war diary anticipates two teams of female nurses for the hospitals at Lychna.) In October 1915, the hospital had 440 beds.

Two views of Field Hospital no. 2 at Lychna. The pier was built by French engineers. After the Gallipoli Campaign, the site was employed by the Royal Naval Air Service.

(*Collection of Bernard de Broglio*)

A large number of works were undertaken in this area by French army engineers between June and December 1915. They built wharves, repaired roads, and developed a water supply for the hospitals and camps around the villages of Lychna and Varos.

In February 1916, the hospital left for Corfu on the hospital ship *Divonna* to care for Serbian troops and refugees, whose winter exodus through the mountains of Albania had left them in wretched condition. The hospital later moved to Florina in Western Macedonia.

Evacuation Hospital (Hôpital d'evacuation) no. 2

This hospital was sited about 500 metres northeast of Field Hospital no. 2, clustered about a small hill with a single windmill, whose ruined base is visible from the Mudros–Myrina road today.

Evacuation Hospital no. 2 at Lychna.
(Ministère de la Culture / APOR 026265)

Work began on huts and facilities in late July, and the hospital is said to have opened on 17 August 1915, with capacity growing from 250 to 500 beds.

By September, the hospital consisted of five barrack huts and 122 tents. Hutted accommodation was built for 12 nurses of the Red Cross, although it is not clear if they joined the hospital before it closed. As well as wooden barracks, buildings in stone were erected, including a mortuary and a facility for cleaning, disinfecting and sterilising. By October, the hospital had 400 beds.

Evacuation Hospital no. 2 was transferred to Mytilene (Lesbos) in January 1916.

Naval quarantine hospital (Lazaret de la marine)

The establishment of an isolation hospital was ordered in June 1915 to treat contagious diseases like typhus, typhoid and cholera. These epidemics had proved devastating during the Crimean and Balkan wars. The *lazaret* was managed by the French naval medical service.

The quarantine hospital appears to have been established on high ground southeast of Meganoros Point, some 2.5 km (1.5 miles) south of Mudros town. A 1917 chart produced by the UK Hydrographic Department notes a 'hospital' at this location. (The published 1920 chart, based on the same survey, amends the label to 'sanitorium'.) The likelihood of this being the *lazaret de la marine* is suggested by the proximity of a pier, 500 metres (1,500 feet) distant, built by the French in 1915. (The British named the landing stage 'Sports Pier'.)

Further evidence is found in a sketch map in Dr Peaudeleu's book, which places the *lazaret* in this general area. It also features in the doctor's description of sunset from the main evacuation hospital, looking across to Meganoros Point: 'Every evening, at dusk, the sky was ablaze with purple and orange strata, the dazzling layers of which contrasted with the dark mass of the hill dominated by the sailors' *lazaret*.'

The war diary of the French engineers notes that the hospital was in the direction of Parthenontos (perhaps modern-day Parthenomytos).

During July and August 1915, French engineers levelled and terraced the ground, and established a 30 cubic metre reservoir. Six Bessonneau huts, 30 metres in length, were erected as wards, a seventh being added later. The facility included an administration hut, laboratory and kitchens. The hospital opened in August or September 1915 with 168 beds.

Structures that might once have formed part of the *lazaret* were documented by Margaret Carter and Steve Moore in *The Gallipolian* journal (no. 136, 2014, 'French Hospital on Limnos').

Officers' rest home (Maison de convalescence)

A four-bedroom rest house for convalescent officers was established in Kastro (Myrina), the principal town on Lemnos island. Capacity was between

10 and 16. The facility, probably administered by Evacuation Hospital no. 1, was opened in September 1915. It closed in February 1916.

Field Hospitals nos. 3 & 4

A third and perhaps fourth field hospital, each with 200 bed capacity, were earmarked for Mudros, but reassigned when the French drew down their forces at Gallipoli in the latter part of 1915.

Military cemetery, Mudros

Those service personnel who died *(Mort pour la France)* on Lemnos island were buried in a military cemetery that was established beside a civilian necropolis, east of Mudros town. The cost of purchasing the land was shared with the British.

Today, East Mudros Military Cemetery is administered by the Commonwealth War Graves Commission. It contains no French war graves.

After the war, the French government intended to consolidate Lemnos burials in Salonika (Thessaloniki). When the Dardanelles veterans' association learnt of this plan, they demanded of the government that Gallipoli fallen be reinterred at Sedd-el-Bahr. The Minister acquiesced. Of 632 French war dead at Mudros, 133 were repatriated to France at the request of their families, and 499 now lie in the French national cemetery on the Gallipoli Peninsula.

But there remains at Mudros a memorial to

French war dead that has a direct connection to the Gallipoli Campaign. French engineers began work on this monument in late December 1915. Stone was quarried and the foundations dug. Craftsmen were sought from French military personnel to dress the stone and cast bronze plaques. The team, comprising some 15 men, hurried to complete the memorial before the Mudros base was shut down, but it was not finished when, at the end of February 1916, the last of the base quit Mudros for Salonika, Corfu and France. A detachment of 3 junior officers and 27 men was left behind to complete the monument that can be seen today: a gabled pedestal atop a broad square plinth that supports a squat obelisk.

In 1930, a French pilgrimage to Gallipoli visited Lemnos. The veterans, informally known as 'Dardas,' placed a bronze plaque on the monument. General d'Amade, the first commander of the *Corps expéditionnaire d'Orient*, led the service. Among those laying wreaths was General Ruef, who had commanded (as a colonel) the landing party at Kum Kale on 25 April 1915. The island of Lemnos was represented by the mayor of Mudros, clergy, Greek troops, and local residents.

Today, the French memorial at East Mudros Military Cemetery bears just two plaques, one from the 1930 pilgrimage, a second dated 14 July 2008. Perhaps those who built the memorial in 1915/16 were unable to produce bronze plaques for the monument, or the 1915/16 plaques have been lost.

The French memorial in Mudros cemetery,
photographed in 1921.

*(Thomas F. Nangle, Memorial University
Archives and Special Collections)*

The French memorial and cemetery, c.1918.

(William Myles Garner, University of Leeds, Liddle Collection)

Grave in Mudros cemetery of *Médecin aide major* (Lieutenant)
Alphonse Benoît Marie Chassy, of Field Ambulance
no. 3, who died of illness on 6 August 1915, aged 46.

Dr Chassy's remains now lie in grave no. 520 (marked Cassy) in
the French national cemetery at Sedd-el-Bahr.

(Collection of Bernard de Broglio)

Hospitals on the island of Tenedos

Citadel Field Hospital (Hôpital de campagne, citadelle)

Field Hospital no. 1 moved from Sedd-el-Bahr castle to the island of Tenedos (modern-day Bozcaada) on 21/22 September 1915. The hospital was established within the castle (originally Venetian, rebuilt in Ottoman times) that stands beside the small harbour of the main town. Space within the castle was soon deemed insufficient, and, on 27 September, a new hospital was ordered to be built under barracks on the island's outhwest coast at Paraskevi (Ayazma).

The hospital in the citadel officially opened on 22 November 1915 but was probably receiving patients before that date. Two buildings within the upper courtyard were probably used. (The ruins of one

'The Red Cross among the isles of Greece: open-air treatment in a British hospital somewhere in the Aegean.'
(The Illustrated War News, 26 April 1916)

Île de Ténédos. L'hôpital de campagne dans la citadelle.
The nurses are British, serving with the French
Red Cross under Mrs Doughty-Wylie.

(Gaston Chérau, Ministère de la Culture /APOR 058434)

remain extant, the other might have been lost
during the extensive renovation that took place
between 1965 and 1970.) Capacity was initially
40, apparently reaching a maximum of 120 beds.
The hospital took patients evacuated from Helles
who could not immediately be received at the
Paraskevi camp, as well as sick and wounded from
the Tenedos garrison.

The hospital had eight army nurses or orderlies.
The staff was supplemented in early November
1915 by 11 'English' nurses under the leadership

of Madame Doughty-Wylie, widow of Charles Doughty-Wylie, who was awarded the Victoria Cross for his part in the landings of 25/26 April 1915 at Helles. Lilian ('Judith') Doughty-Wylie had been anxious to reach the Dardanelles in order to visit the grave of her husband, and she probably travelled to V Beach from Tenedos.

In early January 1916, the hospital moved to Mitylene. Tenedos came under British control, and was garrisoned by troops of the Royal Naval Division (RND). Mrs Doughty-Wylie may have left a small detachment at Tenedos for their care (a press photo dated 1916 shows British soldiers and her nurses in the castle) but at some point she transferred to Mudros, serving in the RND hospital there. In 1917, Mrs Doughty-Wylie established a hospital on the island of Thasos, which she headed until the end of the war.

Paraskevi Field Hospital (Hôpital de campagne, dit d'Aga Paraskévi)

Construction began on 27 September 1915 of a hutted 600-bed hospital at Aga Paraskevi (modern-day Ayazma) on the southwest coast of Tenedos. A lease was signed on 11 November with the Greek community that owned the land.

By December 1915, the hospital was receiving patients. Wounded and sick were landed at a purpose-built hospital wharf from the hospital ship *Saint François d'Assise*. The hospital had reached a

200-bed capacity when Gallipoli was evacuated and the hospital closed. Its patients were transferred to Mudros, medical supplies to Mytilene.

The hutted wards were taken over as barracks by the Royal Naval Division, who garrisoned the Aegean island bases in 1916.

Officers' convalescent home

A Greek mansion near Paraskevi was used as a rest home for officers. It opened on 9 August 1915, with 15 beds. The facility closed when French forces quit Tenedos in early 1916. (The house, now a picturesque ruin, still stands.)

Hospitals at Mytilene (Lesbos)

Evacuation Hospital (Hôpital d'évacuation) no. 2

This hospital was transferred from Lemnos (Lychna) in January 1916, opening on 3 February with a capacity of 300 beds. (The hospital had been preceded by a party to erect huts and barracks.) It was located at Lutra (modern-day Loutra) on the Gulf of Iero (Kolpos Geras).

The hospital served the 1st Division, reconstituted on Mytilene as the 17th Colonial Infantry Division, until its departure for Salonika in March 1916. From this date, the hospital was reduced in both personnel and equipment (transferred to Salonika) and re-named the Lutra-Mytilene Auxiliary Hospital *(hôpital temporaire)*. In June 1916, its capacity was 150 beds.

Amidst olive groves between Loutra and Skala Loutron can be found the site of a French wartime cemetery associated with the hospital. No graves remain, but the plot retains its walls and a weathered memorial stone inscribed with olive branch motifs.

Hospital ships

The French navy supported the expeditionary force with hospital ships. These assumed greater importance when the Allies were held up within their shallow position at Helles, under artillery fire from front and flank. Hospitals could not be landed, so ships functioned as front-line hospitals, receiving wounded from the firing line, after triage ashore. The *Canada*, *Duguay Trouin*, *Tchad* and *Bretagne* (renamed *Bretagne II* from October 1916) were anchored two or three nautical miles off Cape Helles, occasionally exposed to enemy fire. Wounded were brought to the hospital ships in barges and tugs.

On one particular day, *Canada* received 625 seriously wounded in less than 18 hours. She weighed anchor as soon as the last casualty had been brought on board. Emergency dressings and operations continued throughout the crossing.

Medical evacuation was typically to Bizerte or Toulon, the journey lasting from three to five days. During the passage, medical and nursing staff worked all day and sometimes all night.

Canada made 13 such voyages between May 1915

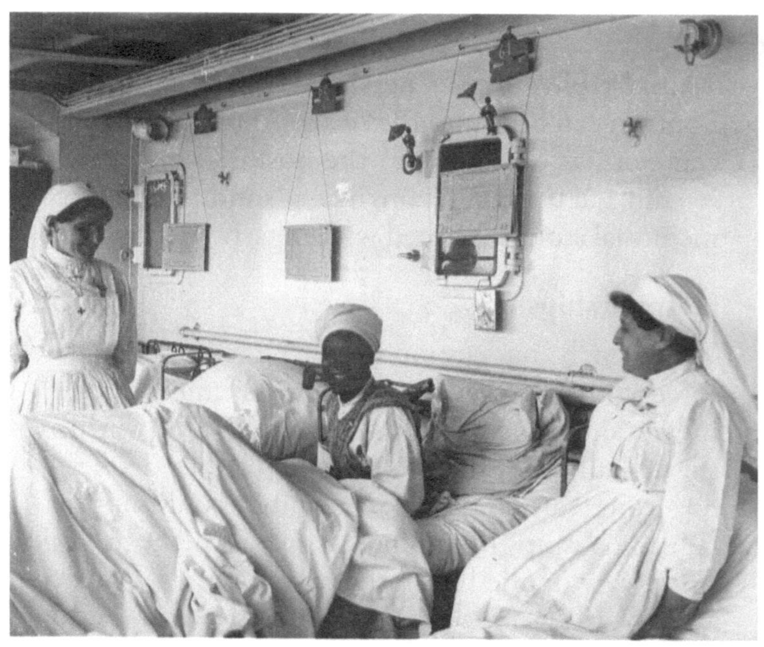

Red Cross nurses with soldier Patinet, 176th Infantry
Regiment, aboard the hospital ship *Charles Roux*.

(Gaston Chérau, Ministère de la Culture / APOR 032522)

Hospital ship *Duguay Trouin* lying off the Gallipoli Peninsula.

(Kerboul, Ministère de la Culture / APOR 020567)

and April 1916, evacuating 7,572 wounded and sick. She operated as a hospital for 328 days, her medical personnel performing 288 operations.

Later in the campaign, as dysentery, jaundice and other afflictions of trench-life took hold, more sick than wounded were evacuated. From July 1915, *Canada* was transporting about two-thirds sick and one-third wounded.

Among the wounded evacuated by hospital ship was CEO commander, General Henri Gouraud. On 30 June 1915, when visiting the field ambulance at Sedd-el-Bahr, the blast of an Ottoman artillery shell hurled him over a wall, his fall broken by a fig tree. Gouraud was taken aboard *Tchad*. During the passage, gangrene broke out, and Gouraud's arm was amputated by Doctor P. Oudard, head of surgery at Toulon.

More ships were requisitioned for medical evacuation when the campaign stalled. These included *Bien Hoa, Ceylan, Charles Roux, Divona, Saint François d'Assise, Sphinx* and *Vinh Long*. They loaded fewer wounded at Cape Helles, but filled up at Mudros, then sailed for France or Algeria.

A team of 10 navy doctors, with nurses and medical equipment, was stationed at Mudros, ready to embark on the transports. No less than eight hospital ships were on station at any one time. The Australian medical history comments favourably on the efficiency of the CEO's chief transport officer at Mudros.

Aboard the hospital ship *Sphinx*, a converted ocean liner.
(Elisabeth Jardin Collection)

Helles seen from the hospital ship *Duguay Trouin*.
Left of picture, a British hospital ship stands off the Gallipoli
Peninsula, the heights of Achi Baba rising behind its aftermast.
(Kerboul, Ministère de la Culture / APOR 020558)

The French Red Cross societies also supported the naval medical service. During the Gallipoli Campaign, the hospital ships *Charles Roux* and *Saint François d'Assise* were funded and equipped by the *Société de secours aux blessés militaires*.

Charles Roux, which reached Mudros on 28 August 1915, made a particularly valuable contribution during the campaign. Two eminent surgeons were amongst a first-class medical team that included 22 female nurses, led by *Mme la marquise* de Clapiers, who would be awarded the *Croix de Guerre* for her and the team's efforts. The nursing sisters included *mesdames* Carteron, Dézaneau, de Marthille, Trousseau, and Blanc, and *mesdemoiselles* Voisin, Lopez, Revol, Prospert, Baduel-d'Oustrac, Ferret, Trespaillé, Eugénie Pietrowska, Dutil, and Batut. Completing the nursing staff were six sisters from the Daughters of Charity of Saint Vincent de Paul.

When the *Charles Roux* could not coal at Mudros, the ship made for Salonika or Athens. On 17 October 1915, when in the Greek capital, the doctors and nurses of *Charles Roux* were honoured with a reception at the French legation, where they were toasted by the pro-Allied Greek statesman, Eleftherios Venizelos.

———

Table: French hospitals of the Gallipoli Campaign

Hospital	Location	Beds	Date start	Date end
GALLIPOLI PENINSULA				
Field Hospital no. 1	Sedd-el-Bahr castle	88	29.04.1915	23.09.1915
Field Ambulance no. 1	Kum Kale; Sedd-el-Bahr castle (former Turkish hospital)	97	11.05.1915	
Field Ambulance no. 2	Sedd-el-Bahr castle (lower courtyard)	190/266	13.06.1915	
Field Ambulance no. 3	Sedd-el-Bahr castle (upper courtyard)	50	22.05.1915	
Field Ambulance no. 4	Entrance to village of Sedd-el-Bahr	74	13.07.1915	
LEMNOS				
Greek hospital	Mudros	20	15.03.1915	24.03.1915
Evacuation Hospital no. 1	Mudros	200/1,500	26.03.1915	15.02.1916
Infirmary	Mudros	150/180	01.06.1915	1918
Convalescent camp	Bay of Pournia		14.06.1915	
Field Hospital no. 2	Lychna	750	15.06.1915	07.02.1916
Evacuation Hospital no. 2	Lychna	250/500	17.08.1915	12.01.1916
Naval quarantine hospital	Meganoros hill	168	[Sept 1915]	[1918]
Officers' rest home	Kastro	10/16	[Sept 1915]	16.02.1916

Citadel Field Hospital [ex-Sedd-el-Bahr]	Tenedos castle	40/120	22.11.1915	Jan. 1916
Paraskevi Field Hospital	Paraskevi	200	07.12.1915	11.01.1916
Officers' convalescent home	Near Paraskevi	15	09.08.1915	[Feb. 1916]

MYTILENE

Evacuation Hospital no. 2 [ex-Mudros]	Lutra	300	03.02.1916	[March 1916]
Auxiliary Hospital [ex-Evacuation Hospital no. 2]	Lutra	150	[March 1916]	08.02.1919

'How we dine at the Sedd-el-Bahr camp while the shells are bursting. May 1915.'

(Elisabeth Jardin Collection)

Mudros scenes.

(Elisabeth Jardin Collection)

IMAGE CREDITS

Photographs are credited throughout the text. The full-page images used as the frontispiece and section dividers are presented without captions to preserve their visual impact; their descriptions are given below. Except for the frontispiece, all are taken from Elisabeth Jardin's album.

Frontispiece: 'French Colonial Infantry disembark at Lemnos.' Press photograph by Maurice Branger.

Part I: On board ship bound for Corfu, January 1916. The verso caption lists the team: Elisabeth Jardin, *Mlle* Niquet, *Mlle* Lacaze, *Mlle* Oberkampf, and *Mme* Godineau.

Part II: Uncaptioned, but similar photos describe a ward in Evacuation Hospital no. 1 at Mudros.

Part III: Mass celebrated on the hill above Evacuation Hospital no. 1.

Part IV: One of a number of photographs of the local environs, collectively captioned *'Le village.'*

Part V: Elisabeth Jardin with camera in hand, photographing a soldier?

Part VI: Christmas, Mudros, 1915.

Part VII: A portrait of Elisabeth Jardin?

Appendix: French infantry in a lighter off Mudros.

BIBLIOGRAPHY

Ministère des Armées, Mémoire des hommes
(memoiredeshommes.sga.defense.gouv.fr),
*Corps expéditionnaire d'Orient, journaux des marches et
des opérations* [war diaries] (26 N 75/1 to 26 N 75/7,
26 N 75/16, 26 N 76/5, 26 N 76/11)

Ministère de la Culture, base Léonore (leonore.archives-
nationales.culture.gouv.fr), *dossiers de la Légion d'honneur*

*Science et dévouement: le service de santé – la Croix-Rouge:
les oeuvres de solidarité du guerre et d'après-guerre*
(Paris: Aristide Quillet, 1918)

*Amicale des infirmières et assistantes sociales de la Croix-Rouge
française,* 'La Croix-Rouge en Orient', *Bulletin trimestriel
de l'Association mutuelle des infirmières de la Société de
secours aux blessés militaires,* no. 13, January 1916

Antelme, Jeanne, *Avec l'armée d'Orient: Notes d'une
infirmière à Moudros* (Paris: Émile-Paul Frères, 1916)

*Association Nationale des groupements d'anciens combattants
des Dardanelles, Les Dardanelles Expédition de 1807,
Expédition de 1915-1916, Pèlerinage de Juin 1930*
(Paris, 1933)

Darrow, Margaret H., 'French Volunteer Nursing and the
Myth of War Experience in World War I', *The American
Historical Review*, volume 101, no. 1, February 1996

Dudgeon, Leonard S., 'Personal Experiences on the
Gallipoli Peninsula and in the Eastern Mediterranean
While a Member of the War Office Committee for
Epidemic Diseases and Sanitation', *Journal of the Royal
Society of Medicine*, vol. 9, July 1916, pp. 101–118

Fabre, Elisabeth (née Jardin), *Un hôpital d'évacuation aux Dardanelles: hôpital d'évacuation n° 1, Moudros (avril 1915 – février 1916)* (Paris: *Faculté de Médecine*, 1920)

Garcia, Julien, Hugues Lefort, Christophe Lamache, Xavier Tabbagh, François Olier, 'Les infirmiers militaires français dans la guerre 1914–1918', *Soins, la revue de référence infirmière*, volume 59, no. 786, June 2014

Jacquet, Benjamin, *Les malades de la Grande Guerre: les poilus et leurs médecins face à la maladie* (Paris: Ellipses, 2021)

Lachèse, Marie-Christine, and Bernard Lachèse, *Un médecin militaire aux colonies: Joseph Briand (1897–1921) Afrique, Asie, Dardanelles* (Paris: L'Harmattan, 2021)

Le Goaer, Charles-Louis, *Rôle de la marine dans l'évacuation des blessés et des malades pendant la dernière guerre (1914–1918)* (Bordeaux: A. Destout, 1919)

Morillon, Marc, and Jean-François Falabrègues, *Le Service de santé 1914–1918* (Paris: Bernard Giovanangeli, 2014)

Olier, François and Jean-Luc Quenec'hdu, *Hôpitaux militaires dans la guerre 1914–1918*, vol. 5, *Front du Nord-Est & armée d'Orient* (Louviers: Ysec, 2016)

Peaudeleu, Dr. Léon, *Aux Dardanelles, à Lemnos, sur les bords du Vardar: Souvenirs de la Guerre d'Orient et Impressions de Voyage* (Nice: Imprimerie du Patronage Saint-Pierre, 1920)

Sardet, M., 'La Marine et ses navires-hôpitaux dans les Flandres et en Orient', *Médecine et armées*, vol. 44, no. 1, February 2016

N
AUSTRIA-HUNGARY
BALKANS &
AEGEAN SEA
ROMANIA
Belgrade
Bucharest
SERBIA
Danube R.
MONTE-
NEGRO
BULGARIA
BLACK
SEA
Sofia
Maritza R.
ALBANIA
Constantinople
MARMARA SEA
MACEDONIA
Salonika
Mt Athos
Chanak
Bithynian
Olympus
Mt
Olympus
Mt Ida
CORFU
GREECE
AEGEAN SEA
TURKEY
Athens
Smyrna
Key
DODECANESE
(It.)
GALLIPOLI
PENINSULA
LEMNOS
CRETE
SKYROS
MYTILENE
MEDITERRANEAN
SEA
LIBYA
(It.)
Alexandria
EGYPT
(Br.)

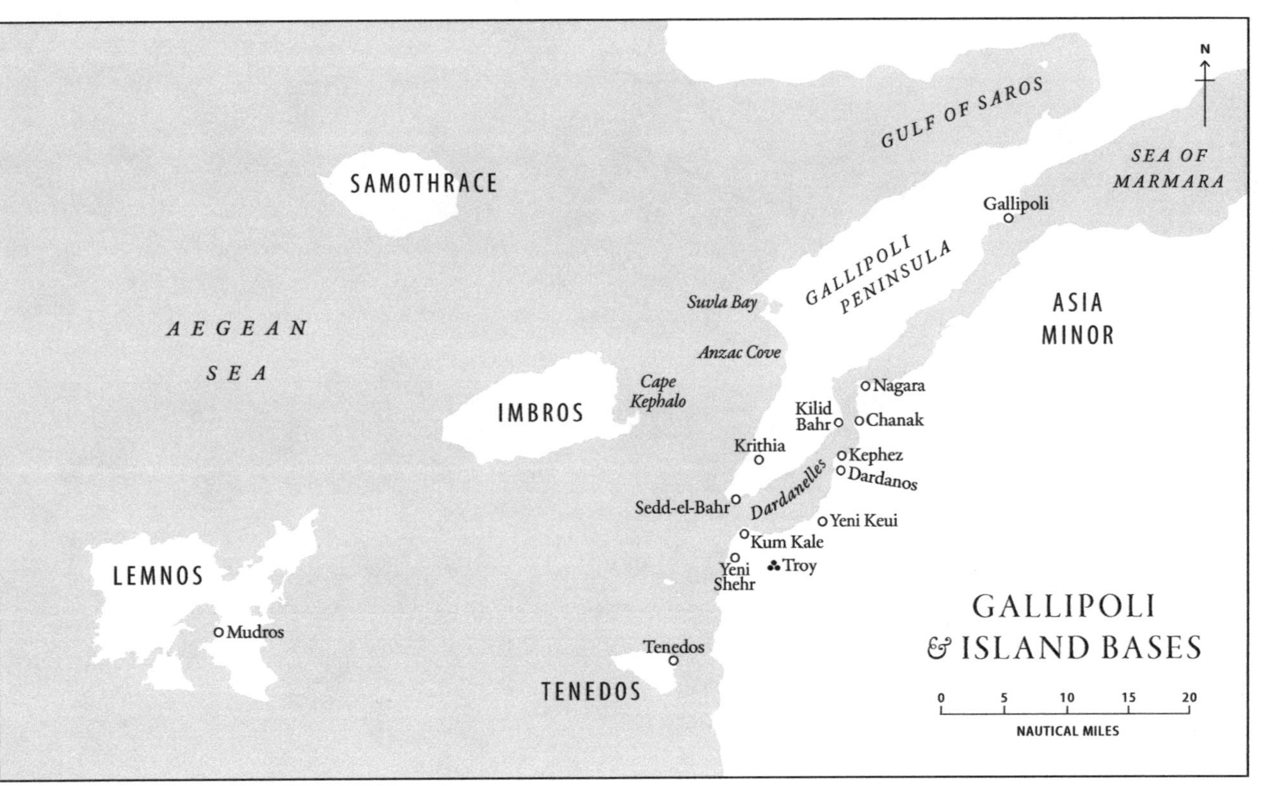

N
GULF OF SAROS
SEA OF MARMARA
SAMOTHRACE
Gallipoli
GALLIPOLI PENINSULA
ASIA MINOR
Suvla Bay
AEGEAN SEA
Anzac Cove
Nagara
Cape Kephalo
IMBROS
Kilid Bahr
Chanak
Krithia
Kephez
Dardanos
Dardanelles
Sedd-el-Bahr
Yeni Keui
Kum Kale
Yeni Shehr
Troy
LEMNOS
Mudros
Tenedos
TENEDOS
GALLIPOLI & ISLAND BASES
0 5 10 15 20
NAUTICAL MILES

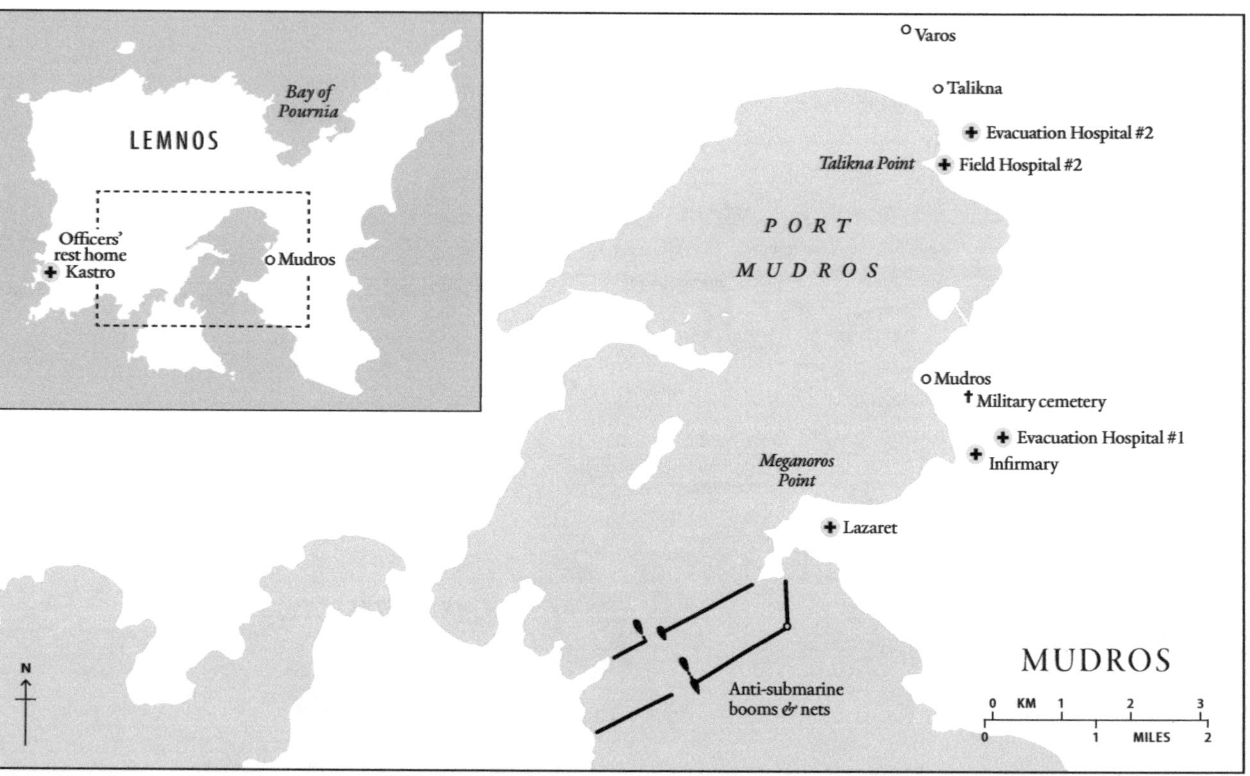

LEMNOS
Bay of Pournia
Officers' rest home
Kastro
Mudros
Varos
Talikna
Evacuation Hospital #2
Talikna Point
Field Hospital #2
PORT
MUDROS
Mudros
Military cemetery
Evacuation Hospital #1
Infirmary
Meganoros Point
Lazaret
Anti-submarine booms & nets
MUDROS
N
KM
0   1   2   3
MILES
0   1   2

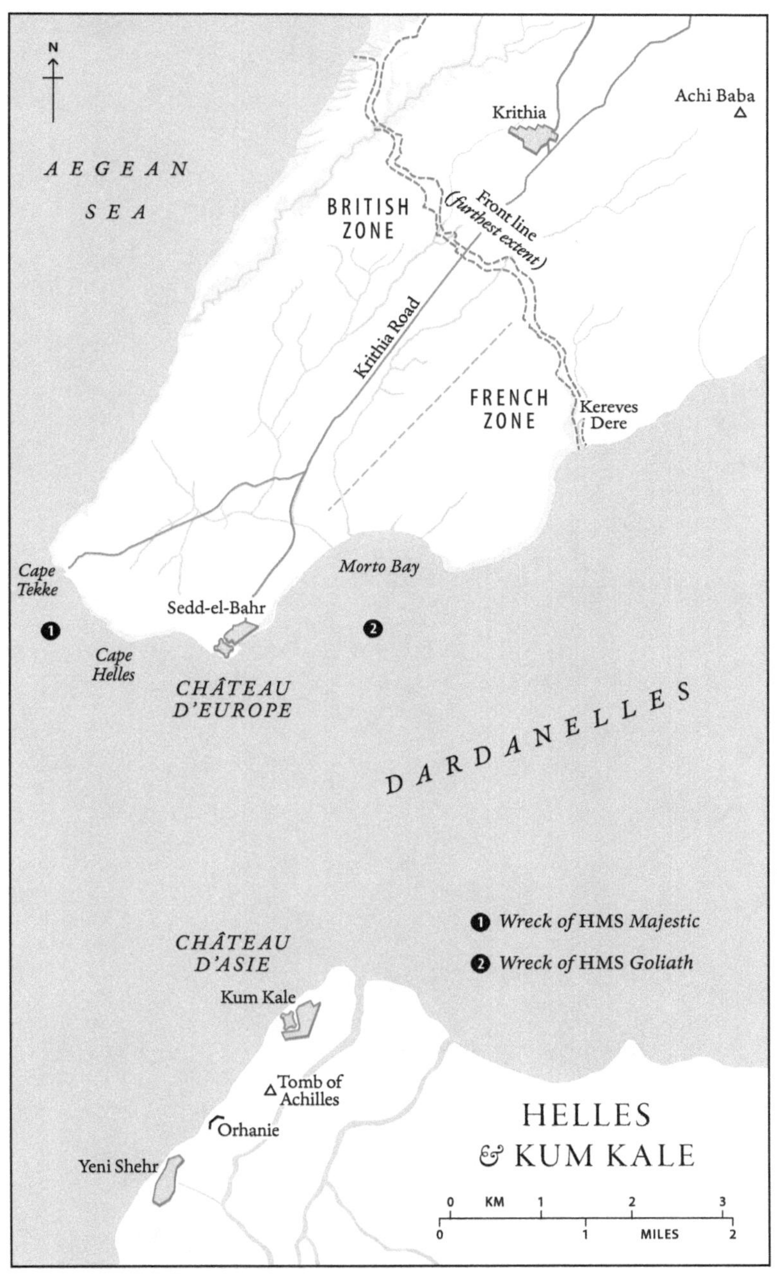

N
AEGEAN SEA
Krithia
Achi Baba
BRITISH ZONE
Front line (furthest extent)
Krithia Road
FRENCH ZONE
Kereves Dere
Cape Tekke
Morto Bay
Sedd-el-Bahr
Cape Helles
CHÂTEAU D'EUROPE
DARDANELLES
CHÂTEAU D'ASIE
Wreck of HMS Majestic
Wreck of HMS Goliath
Kum Kale
Tomb of Achilles
Orhanie
Yeni Shehr
HELLES & KUM KALE
0   KM   1   2   3
0   1   MILES   2